ABHANDLUNGEN DER DEUTSCHEN AKADEMIE DER WISSENSCHAFTEN ZU BERLIN

Klasse für Medizin

Jahrgang 1960 Nr. 1

CHEMISCHE MUTAGENESE

ERWIN-BAUR-GEDÄCHTNISVORLESUNGEN I 1959

Veranstaltet vom Institut für Kulturpflanzenforschung Gatersleben der Deutschen Akademie der Wissenschaften zu Berlin

26.–28. Juli 1959

Herausgegeben von

H. STUBBE

Redaktion

H. BÖHME

Mit 44 Abbildungen, davon 8 Abbildungen auf 3 Kunstdrucktafeln, und 15 Tabellen

AKADEMIE-VERLAG · BERLIN

1960

Vorgelegt von Hrn. Stubbe in der Gesamtsitzung vom 19. November 1959
Zum Druck genehmigt am gleichen Tage, ausgegeben am 10. März 1960

Erschienen im Akademie-Verlag GmbH, Berlin W 1, Leipziger Straße 3-4

Lizenz-Nr. 202 · 100/701/60
Gesamtherstellung: IV/2/14 · VEB Werkdruck Gräfenhainichen · 1257
Bestellnummer: 2001/60/IV/1
Preis: DM 18,50
Printed in Germany
ES 18 G 1

INHALTSVERZEICHNIS

ERWIN BAUR (1875–1933)

Einleitende Worte

Am 2. Dezember 1958 fand im Institut für Acker- und Pflanzenbau Müncheberg der Deutschen Akademie der Landwirtschaftswissenschaften zu Berlin, dem ehemaligen Kaiser-Wilhelm-Institut für Züchtungsforschung in Müncheberg/Mark, eine Gedächtnisfeier für Erwin Baur statt aus Anlaß der 25. Wiederkehr seines Todestages. Als wir die Feier vorbereiteten, überlegten wir, wie wir das Andenken an diesen großen deutschen Vererbungsforscher der ersten Jahrzehnte unseres Jahrhunderts wachhalten könnten, damit er einer jungen Generation von Genetikern stets Vorbild bleibe.

Zwei Wege sind nun beschritten worden, um Erwin Baur ein Vierteljahrhundert nach seinem Tode zu ehren.

Die Deutsche Akademie der Landwirtschaftswissenschaften zu Berlin hat das Statut zur Verleihung einer Erwin-Baur-Medaille vorbereitet, die für besondere Verdienste auf dem Gebiet der Landwirtschaftswissenschaften alle zwei Jahre auf den Festsitzungen der Akademie verliehen werden soll.

Im Institut für Kulturpflanzenforschung Gatersleben der Deutschen Akademie der Wissenschaften zu Berlin haben wir beschlossen, im Rahmen der Symposien unseres Instituts alle zwei bis drei Jahre „Erwin-Baur-Gedächtnisvorlesungen" abzuhalten, wobei zusammenfassenden Darstellungen über bestimmte, intensiv bearbeitete Gebiete der Genetik möglichst gründliche wissenschaftliche Gespräche von Fachleuten in kleinem Kreise folgen sollen.

Wir hoffen, daß diese Zusammenkünfte allen Beteiligten von Nutzen sein werden, und wir wünschten uns, daß bei solchen Gelegenheiten auch Kollegen einander treffen, die sich aus ihren Arbeiten zwar kennen, die sich aber noch nicht begegnet sind. Über den strengen Austausch wissenschaftlicher Gedanken hinaus erhoffen wir aber auch persönliche Gespräche in einer freundlichen und freundschaftlichen Atmosphäre, die immer wieder zeigen werden, wie leicht und beispielhaft die gute Verständigung der Wissenschaftler aller Nationen in unserer Welt ist.

Unsere ersten „Erwin-Baur-Gedächtnisvorlesungen" stehen unter dem Thema „Chemische Mutagenese", ein Gebiet, in dem in den letzten Jahren ungeahnte Fortschritte gemacht wurden. Die alten Pioniere unserer Wissenschaft, darunter auch Erwin Baur, haben schon zu Beginn dieses Jahrhunderts die willkürliche Beeinflussung der Erbsubstanz durch äußere Einwirkungen in Betracht gezogen und die ersten Versuche an niederen Organismen, auch durch chemische Beeinflussung Mutationen zu erzeugen, angestellt. Diese Untersuchungen sehen heute vom Standpunkt einer verfeinerten Methodik, einer zweckmäßigen Wahl der Versuchsobjekte, einer engen Verbindung zwischen Biochemie und Genetik sehr primitiv aus. Das hat seinen Grund in zweierlei: einmal arbeiteten diese Genetiker meist allein, ohne die Mitwirkung von Spezialisten benachbarter Disziplinen, und zum anderen waren die benachbarten Disziplinen selbst noch nicht so weit in ihren Erkenntnissen fortgeschritten, daß sie Ansatzpunkte zur Gemeinschaftsarbeit sahen. Dennoch müssen wir dieser alten Generation von Genetikern zuerkennen, daß sie eines der Hauptgebiete neuzeitlicher Genetik schon damals in ihren Grundzügen erkannt, wenn auch längst nicht einer befriedigenden Lösung zugeführt hat. Man gewinnt dabei immer wieder den Eindruck, daß die Lösung eines Problems einer langen Vorbereitungszeit bedarf, bis es reif wird für eine großartige und zu neuen Erkenntnissen führende Entwicklung des Fachgebietes. Auf diese Weise wird, wenn die ersten Erfolge zu verzeichnen sind, ein Teilgebiet einer Wissenschaft plötzlich „modern", und die Kurve der Zahl der Bearbeiter in allen Ländern und damit auch die Zahl der Ergebnisse steigt steil an.

Es ist für unsere Zeit charakteristisch, daß intensive Fortschritte fast nur in der Gemeinschaftsarbeit benachbarter Disziplinen zu erreichen sind, und unsere Zusammenkunft beweist ja, daß wir auf die Zusammenarbeit mit Chemikern und Physikern in der Genetik nicht mehr verzichten können. In dieser team-Arbeit sind manche Länder inzwischen vorbildlich geworden, in Deutschland muß jedoch manches verbessert werden, da hin und wieder psychologische Gründe die Gemeinschaftsarbeit erschweren wenn nicht verhindern. Dennoch sind überall erfreuliche Ansätze vorhanden, da sich die Erkenntnis Bahn bricht, daß niemand mehr allein, aus der Sicht seines mehr oder weniger engen Fachgebietes wirklich grundsätzlich Neues erarbeiten kann.

So ist zu hoffen, daß die „Erwin-Baur-Gedächtnisvorlesungen" des Jahres 1959 dazu dienen werden, das bisher auf dem Gebiet der chemischen Mutagenese an Tieren, höheren Pflanzen und niederen Organismen Erarbeitete zusammenfassend darzustellen und die Problematik aufzudecken, daß in den Diskussionsbeiträgen die letzten Ergebnisse der Forschung von einer Anzahl namhafter Spezialisten dargelegt werden und neue Richtungen der Forschung und vielleicht auch der Zusammenarbeit erkennbar werden.

H. Stubbe

Chemical mutagenesis in animals

C. Auerbach

Institute of Animal Genetics, Edinburgh

It is now almost 20 years since highly effective chemical mutagens have become available for study. We are most grateful to Professor Stubbe and the Deutsche Akademie der Wissenschaften for giving us this splendid opportunity to take stock of past achievements and to coordinate plans for future research in the field of chemical mutagenesis. To me has been allotted the task of speaking on the work done with animals. Undoubtedly, this does not include the most exciting results of recent years; but to some extent, this shortcoming is compensated for by the variety of problems which have been studied on animals. Indeed, this variety is such that I shall find it difficult to make coherent whole of my talk, even if I restrict myself to what to me seem the most important lines of approach. Necessarily, the selection of quoted authors and results will be quite incomplete, and I must apologise in advance for this unavoidable discourtesy.

I. Screening of substances for mutagenic ability

(1) *Drosophila* lethals

A large proportion of this type of work has been done on *Drosophila*, using the frequency of recessive sex-linked lethals as a measure of mutagenic success. For two reasons, this method has recently come under fire from the camp of micro-organism genetics. The first objection is that lethals are a mixed population, consisting of point mutations and small and large re-arrangements in, usually, unknown proportions. This is certainly true, and for valid quantitative comparisons between mutagenic treatments overall mutation frequencies have to be broken down into these various components. The fact that, to a considerable extent, this can be done by cytological methods, turns this apparent shortcoming of lethals into an advantage by providing a tool for a more complete analysis of mutagens. Moreover, there are good grounds for believing that not all lethals are due to deficiencies or position effects, so that even substances with little or no ability to cause chromosome breakage will be able to induce *Drosophila* lethals. The opposite is not necessarily true: substances which can produce re-arrangements but no point mutations, would appear as mutagens in *Drosophila* tests; but no such substances have so far been found. The second objection to the use of lethals is that they occur at an unknown number of loci, which may differ from each other in their response to a given mutagen. Recent observations on mutagen specificity make it very likely that such differences exist. But it seems to me that this possibility, far from disqualifying lethals as a screening agent for mutagens, makes them eminently suitable for this purpose. The use of specific loci for screening tests may give quite misleading results. I only want to remind you of the so-called "mutagen-stable" loci in bacteria. I have little doubt that with suitable treatment – perhaps by purine or pyrimidine analogues – these loci can be made to mutate; but had they been used in screening tests for mutagenic ability, the mutagenicity of X-rays, ultra-violet radiation, nitrogen mustard, and other highly effective chemicals would have remained undetected. In tests for lethals, differences between loci will even out in the average. Moreover, since lethals are "forward mutations" at, presumably, complex loci, compensation between different degrees of sensitivity is expected to occur within as well as between loci.

It is true that for the highly important class of mutagenic purines and pyrimidines lethal tests on *Drosophila* have so far yielded negative or very weakly positive results; but it seems most unlikely that this should be due to specificities which preclude action on any of the loci that can mutate to lethals. It is much more likely that the mutagenic action of these substances depends on biochemical conditions which were not fulfilled in the previous experiments on *Drosophila*, and which, perhaps, cannot be realized at all in such experiments. That the action of a chemical mutagen may be profoundly affected by metabolism and other physiological factors has been clearly shown in experiments in which formaldehyde or pyronin-B were used as mutagens in the food of *Drosophila* larvae. Under these conditions, the same concentration of formaldehyde can produce any value between 0 and 15% sex-linked lethals, depending on the sex of the larvae, the developmental stage of the treated germ cells, the type of food medium used, the number of larvae per food vial, etc. For pyronin, the situation is different. Here, the effectiveness of a given dose depends to a remarkable degree on unknown differences between the physiological states of the individual larvae or between their micro-environments within the same food vial. Thus, in one particular experiment the overall frequency of autosomal lethals was only 3%, but all lethals occurred in the progeny of a few larvae and the most drastically affected animal had no fewer than 8 mutations, 5 of them different, in a total of 20 tested chromosomes.

(2) Effective mutagens

Let us now briefly consider which substances have shown themselves mutagenic in tests on *Drosophila*. Their number is very large, and I shall therefore divide them into broad groups. The largest group consists of the so-called "alkylating agents"; it contains sulphur and nitrogen mustards, epoxides, ethylene imines, and methanesulphonates. It also contains the beta-lactones which so far have not been tested on *Drosophila*, but one of which, beta-propiolactone, is a very effective mutagen for plants and microorganisms. A great variety of alkylating agents has been synthesized at the Chester Beatty Institute in London and tested for mutagenic ability by the FAHMYS. The effective ones are characterized by a diad or triad of effects: they are mutagenic, carcinostatic and, often, carcinogenic. The close correlation between the first two effects is not surprising; for most of these substances were tested for mutagenic ability because they had already shown carcinostatic action which – at least in the case of alkylating chemicals – is mainly due to chromosome breakage. The connection between mutagenic and carcinogenic effects is less obvious; I shall come back to it later on.

Urethane which, with mustard gas, started off the series of effective mutagens, has the same triad of effects as the alkylating agents, but its action appears to be less direct.

Formaldehyde produces mutations when mixed with the food of the larvae or when injected into adult ♂♂. The mechanism of mutagenesis seems to be different for these two types of treatment. There are several reasons for inferring that formaldehyde injection produces mutations via the formation of mutagenic peroxides. Thus, in *Neurospora*, the mutagenic effect of formaldehyde on conidia is vastly enhanced by the addition of hydrogen peroxide, which by itself is very weakly mutagenic. In *Drosophila*, this kind of combined treatment is not possible because of the high catalase content of the body fluids; but Sobels could increase the frequency of formaldehyde-induced mutations by pretreatment of the flies with sodium azide or potassium cyanide. Under these conditions, some mutations could be obtained even in females, which otherwise are completely refractory to the mutagenic action of formaldehyde. Finally, Sobels could show that the addition compound of formaldehyde and hydrogen peroxide, dihydroxy-dimethyl peroxide, produces mutations in the same germ cells which are also responsive to formaldehyde injections. Formaldehyde, given as injection, thus belongs into the same group of chemical mutagens as other peroxides which produce mutations in *Drosophila* and micro-organisms. Formaldehyde mixed with the food of the larvae appears to act in a different way, perhaps by interfering with some synthetic process which is involved in gene replication. It appears to act exclusively during the long growth stage of the primary spermatocyte in the larva, and its action is depressed by a variety of conditions which delay development. This does not seem to be due to a retarded appearance of sensitive cells, but to a dependence of the mutagenic effect on rate and, perhaps, type of metabolism.

Most of you will know that the mutagenic action of formaldehyde food was discovered by Rapoport. It is perhaps less well known that Rapoport also discovered a variety of other mutagens, which subsequently were detected independently by Western geneticists. Of particular interest is the fact that ethylene imine was highly mutagenic in his experiments. This substance has so far not been retested on *Drosophila*, but has given very good results in plants. Its interest lies in the fact that it is a monofunctional compound, and that such compounds in general have much lower carcinostatic abilities than bifunctional ones. Some years ago, this difference between mono- and bifunctional compounds was explained on the basis of the *cross-linkage hypothesis*, which assumed that cross-linkage between biological molecules is a prerequisite for carcinostatic as well as mutagenic ability. For the production of gene mutations, this hypothesis could easily be disproved by experiments on *Neurospora*, where certain monofunctional mustards produced as many or more reverse mutations than bifunctional ones. Chromosome breaks, too, can be produced by monofunctional compounds, e. g. by beta-propiolactone. However, the close connection between carcinostatic and chromosome-breaking abilities suggested the possibility that monofunctional compounds may be less effective than bifunctional ones in producing intergenic changes. If this were true, the ratio of translocations to lethals should be higher for bifunctional than for monofunctional compounds, and this has just been tested by Professor Nakao at our Institute. The two substances which he compared were epoxide and diepoxybutane; contrary to expectation, the ratio of translocations to lethals was at least as high after treatment with the monofunctional as with the bifunctional substance. Unfortunately, epoxide is altogether a weak mutagen for *Drosophila*. Professor Nakao intends to test ethylene imine as a more suitable monofunctional compound.

The selection of chemicals for tests of mutagenic ability has been somewhat biassed by the close connection of this work with cancer research. Whole groups of chemicals have remained unexplored, and some of these may well contain highly effective mutagens. An example is the recent discovery, by Clark in Australia, of the strong mutagenic action of an alkaloid, heliotrin. This brings *Drosophila* in line with plants, in which Oehlkers and his collaborators have produced genetical effects by a variety of alkaloids.

Thymonucleic acid, too, was effective in Oehlkers' experiments, and Gershenson has repeatedly claimed that it produces mutations when mixed with the food of *Drosophila* larvae. Unfortunately, the evidence is far from clear. The overall frequency of mutations is very low: in the most recent experiments, 1 visible mutation occurred per 6000 treated germ cells. Rapoport and other workers had failed to produce sex-linked lethals by similar treatment. Gershenson interprets this to mean that only viable mutations are produced; but the data which he adduces in support of this are not convincing. A priori, it is difficult to imagine that any mutagen should be able to discriminate between viable and lethal point mutations; it is even more hard to imagine a mutagen which, as Gershenson claims for thymonucleic acid, preferentially produces mutations which affect the wings.

(3) Tests for further mutagens

The question may well be asked: Should we go on testing more substances for mutagenic ability? I believe there are good reasons for an affirmative answer. It is now clear that there exists no one magic group which confers mutagenic ability on a substance. On the contrary, the most diverse chemicals can be equally effective mutagens. It seems to me that this situation, instead of discouraging further search for mutagens, should stimulate it; for the diversity of means by which mutations can be produced must ultimately derive from the diversity of mechanisms which maintain stability of the gene and fidelity of gene replication, and a study of the first diversity may well help understand the second. Of special interest in this respect are cases of mutagen-antimutagen balance, such as has been found between certain purines and purine-ribosides in bacteria, and between catalase and formaldehyde in *Drosophila*. Tests of new compounds are, moreover, often useful for following-up a hypothesis on the chemical action of a particular mutagen or group of mutagens. Thus, Sobels' hypothesis on the way formaldehyde injection acts on the germ cells provoked the successful test of dihydroxydimethyl peroxide, and many new alkylating mutagens have been synthesized on the basis of theoretical considerations. In particular the interpretation of mutagen specificity can gain from the testing of new compounds. Suppose it were found that nitrous acid produces a specific pattern of mutations in, say, bacteria. Evidence that a similar pattern is produced by other compounds which act on the amino groups of DNA would then be strong, albeit indirect support for the hypothesis that nitrous acid produces mutations by reacting with these groups.

Finally, as more and more chemicals are used in therapeutics, food processing and other industries, the testing of these substances for mutagenic ability will become a necessary protective measure. The time may also come when tests of chemicals as possible mutagens for economically important animals may be considered of practical value. In either case, results obtained on *Drosophila* are of very limited value, and there is an urgent need for the development of techniques for mutation tests on small laboratory mammals.

(4) Mutation tests on mammals

Techniques for scoring sex-linked or autosomal recessive lethals in the mouse have been tried out in our Institute and in the Atomic Energy Establishment at Harwell. In either case, the result was disappointing even after strong X-ray treatment, presumably because the part of the genome which can be screened by these techniques forms too small a proportion of the whole. The specific loci technique, which has been so successfully used by the RUSSELLS in U.S.A. requires more mice than can be handled in any but the largest and best-endowed research institutes. One is, therefore, left with the scoring of dominant lethals and translocations, and with the incidental observation of visible mutations.

In our Institute, two chemicals have so far been tested on mice: nitrogen mustard (HN-2) by Dr. FALCONER and myself, and triethylenemelamine (TEM) by Mr. CATTANACH. Both are strong alkylating agents, carcinostatic, carcinogenic, and mutagenic for *Drosophila* and plants. Both were applied as intraperitoneal injections into males. TEM proved exceedingly effective; it produced at least as many dominant lethals and translocations as the highest tolerated X-ray dose, and two visible mutations of a somewhat unusual nature. Nitrogen mustard, too, showed undoubted ability to produce translocations in mice; but quantitatively its action was weak, and the use of higher doses was precluded by the strong systemic toxicity of this substance. It may be possible to obtain better results by artificial insemination with treated spermatozoa, and this method may in general be preferable to the more indirect one of injection into the animal.

Recently, methods have become available for the study of mutation in somatic cells of mammals. Individual mammalian cells, grown into clones in tissue culture, can be treated and screened for mutation by the familiar techniques of micro-organism genetics. So far this has been applied only to the detection of drug resistant mutants in populations of human cells. From the applied point of view it may be desirable to use human somatic cells for the tests of substances which may be harmful to the human germ cells. For a more refined genetical study, it would be preferable to work with mouse cells, which can be genetically manipulated in the donor strain.

A third and, in my opinion, very promising approach to the study of mutation in mammalian cells uses mouse tumour cells as material to be treated, and mice as test tubes in which the treated cells can be grown. Again, techniques which are used in mutation research on micro-organisms can be modified to suit the conditions of such experiments. One great advantage of this system over the previously described one is that changes can be studied at the genetically well known loci which determine histocompatibility. In the technique developed by the KLEINS and now used in our Institute by Mr. DHALIWAL tumours are induced in hybrids between two strains which are isogenic apart from one allelic difference at a histocompatibility locus. The hybrid tumours, possessing both antigens, will not take in either parental strain, unless they have lost one or both antigenic specificities. Tumours which have lost both specificities are of little interest genetically; loss of only one specificity may be due to specific cytoplasmic adaptation or to a genetical change. Support for the second alternative consists of the observation that so far such specific variants have occurred only in heterozygotes, and that their frequency can be enhanced by treatment with mutagenic agents. Two such agents, X-rays and TEM, have been tested with positive result. Whether the postulated nuclear event is mutation proper, chromosome loss, deficiency, or somatic crossing-over, cannot be decided on the present evidence. The only clear proof for point mutation would be the acquisition of a new antigenic specificity in place of the lost one; but evidence for this will be difficult to obtain. A different kind of evidence might be provided by showing that two alleles in the same type of hybrid react differently to different mutagens; but the existence of such a high degree of mutagen specificity is doubtful.

Mutation tests on mammalian cells, especially those on somatic cells, may be a good approach to the old problem of whether cancer is caused by somatic mutation. So far the evidence for this theory rests on the observation that all strong and generally effective mutagens – X-rays, UV, nitrogen mustard, and other alkylating agents – are carcinogenic. Against this are adduced the negative results of mutation experiments with the highly carcinogenic polycyclic hydrocarbons. However, these experiments have been carried out on *Drosophila*, and there is no reason for assuming that the metabolic fate of these chemicals is the same in *Drosophila* and mammals. Much greater weight could be attributed to results of experiments in which one and the same substance was used as mutagen for somatic cells and as carcinogen for the donor animal. Such experiments would be of special interest in the case of urethane, a substance which has puzzled geneticists by its ability to produce mutations in *Drosophila*, plants and bacteria, but not in *Neurospora*. In its carcinogenic action, urethane shows an even stricter organism specificity; for it produces lung tumours in mice, but not in rabbits or guinea-pigs. This, as shown by ROGERS in U.S.A., is in part due to the fact that the active carcinogenic agent is a derivative of urethane, which is not formed in all species of rodent. Mutagenic action of this metabolite on mammalian cells would strengthen the somatic mutation theory of cancer.

II. Analysis of the mutagenic process

Mutagenesis is a chain of processes which has many links. The mutagenic agent has to penetrate into the cell. It may have to form a mutagenic derivative. This, or the original agent, must reach the DNA of the chromosome, and in this it may be helped or inhibited by the DNA-associated protein. The first result of interaction between mutagen and DNA often is unstable, requiring stabilization before the replicate gene string has been formed. In chemical mutagenesis, a metastable condition may persist for many cell generations. The newly mutated gene has to produce a mutated organism or clone of cells by altering the biochemical pattern of the cell, and this may have to be done in the face of adverse selection. A detectable mutation appears only after all these processes have been carried to completion. Most of them can be studied better in micro-organisms than in animals. This is true even for delayed mutation, which was first found and analysed in *Drosophila*.

(1) Rearrangements

There exists, however, an additional means of producing changes in the genetic material; this is the production of rearrangements which at least in all higher organisms contribute a large share to the genetical effects of mutagens. For a combined cytological and genetical analysis of rearrangements, *Drosophila* is the material par excellence. I want to discuss briefly four aspects of chemical mutagenesis which have been studied by these means.

(a) From the very beginning of my work with mustard gas I asked myself whether this substance acts as a "hit-poison" in JORDAN's sense of the word, i. e. whether it produces mutations and chromosome breaks by separate, random, and independent events which, in the case of breaks, may subsequently rejoin into rearrangements. The hit nature of X-rays and the breakage-first mechanism by which it produces rearrangements have been inferred from dose-effect curves. For chemical mutagens, dose-effect curves are of little value, at least in higher organisms where the result may be distorted by differences in penetration, interaction with cytoplasmic constituents and other secondary factors. There is, however, one means of overcoming this difficulty. If it may be assumed that the majority of sex-linked lethals are point changes rather than two-break rearrangements – and for mustard gas this could be shown to be so – then the frequency of sex-linked lethals may be used as a measure of the effective mutagen dose, against which the frequencies of rearrangements can be plotted. This experiment has been carried out by NASRAT, and the result was quite unambiguous: within a dose range from 5–12 % sex-linked lethals, translocation frequency rose almost exactly as the square of the dose. This shows that after mustard gas treatment, as after X-radiation, the frequency of breaks increases with dose at the same rate as the frequency of point mutations, and that rearrangements are formed by the union of independently broken chromosome ends.

(b) OSTER extended this work to a study of reunion between breaks which had been produced by X-rays on the one hand, mustard gas or urethane on the other. He treated *Drosophila* males with X-rays followed by mustard gas; mustard gas followed by X-rays; and urethane followed by X-rays. In all cases the frequency of translocations after treatment with two mutagens was higher than additive and agreed well with the assumptions that breaks produced by the first treatment were still open at the time of the second, and that breaks produced by either treatment joined indiscriminately with each other. In addition to once more confirming the breakage-first mechanism for chemically induced rearrangements, these results show that chemically produced breaks have no special features which incline them towards preferential reunion with breaks produced by the same mutagen.

It is interesting to contrast these results with findings on *Vicia* obtained in U.S.A. by WOLFF and his collaborators. They applied X-rays and fast neutrons in quick succession and found that the frequency of rearrangements was additive, showing that breaks produced by either treatment tend to rejoin with similarly produced breaks. They interpreted this to mean that two breaks in the path of the same neutron are usually closer together than either of them is to an independently produced X-ray break. It may be concluded that chemically induced breaks, unlike those produced by neutrons, are spread more or less evenly throughout the nucleus in the same way as X-ray induced breaks. It may be objected that these results may not be true for *Drosophila*, where breaks which have occurred in spermatozoa rejoin only after fertilization. However, the fact that in *Drosophila*, as in *Vicia*, neutrons produce rearrangements by a single-hit mechanism indicates that proximity relations between the chromosomes are not greatly changed when the spermatozoon enters the ovum.

(c) KAPLAN, for mustard gas, and FAHMY and FAHMY, for TEM, found that breaks induced by these chemicals have a greater tendency to remain open than X-rayed induced ones. This was inferred from the fact that dominant lethals, which in *Drosophila* are mainly due to unjoined chromosome breaks, are relatively more frequent after chemical treatment than after irradiation. CATTANACH found the same for mice, where the ratio of dominant lethals to translocations was considerably higher after treatment with TEM than after X-radiation. This peculiarity of chemically induced breaks must be one of the factors which contribute to the relative paucity of large rearrangements after chemical treatment.

(d) Dr. SLIZYNSKA undertook a very careful analysis of the cytological changes in the salivary glands of larvae whose fathers had been reared on formaldehyde food. For an interpretation of her findings it is necessary to keep in mind that all these changes were originally induced in premeiotic germ cells. Her results confirmed those obtained previously by genetical methods: mosaics were frequent, large rearrangements were rare, and the ratio of translocations to inversions was lower than after X-radiation. In addition, she found a very high frequency of deletions and duplications, often in the form of complementary duplication-deficiency mosaics. Many years ago, FORD had found that nitrogen mustard produces many duplications in *Vicia* chromosomes, and he had suggested that these duplications arise through nonhomologous exchanges between sister chromatids. However in *Drosophila*, where the pattern of banding allows a more refined analysis, quite a few duplications turned out to be reverse repeats, and this confirms the breakage-first mechanism of mustard gas induced structural changes. In any case, duplications and their complementary deficiencies are chromatid effects, and their existence in chromosomes which had been treated before meiosis provides further proof for the delayed effect of chemical treatment. The conclusions which Dr. Slizynska draws from her cytological analysis agree with those to which I had been led by genetical data. We consider that delayed opening of breaks is the main cause for the relative shortage of large rearrangements after chemical treatment, and we think that the prevalence of intrachromosomal changes may be due to a tendency for latent breaks in the same chromosome to open simultaneously. As the underlying cause of this tendency Dr. Slizynska suggests a connection between the opening of latent breaks and some particular step in chromosome replication, which in different chromosomes may occur at different times. For the sex-chromosomes, lack of synchrony with the autosomes is a well-known fact, and Dr. Slyzinska's hypothesis derives some support from the finding that in her experiment the X-chromosome had the highest ratio of intrachromosomal to interchromosomal changes.

(2) Brood pattern analysis

This term has been applied to a technique for inferring differences in mutagen sensitivity between germ cells from a comparison of mutation frequencies in successive broods from treated males. This technique has become very fashionable among *Drosophila* workers and is indeed indispensable for quantitative comparisons between mutagens. For the analysis of mutagenic processes its value is limited. It is true that brood pattern analysis of X-ray induced mutations has clearly established a peak of sensitivity in the spermatids of *Drosophila* and the mouse; but the interpretation of brood patterns obtained with chemical mutagens is complicated by the possibility that different germ cell stages may differ in permeability and/or chemical reactions between mutagen and cytoplasm. Similarity and dissimilarity of brood patterns between chemical mutagens does not therefore point unambiguously to similarity or dissimilarity of mutagenic processes, although very marked differences in pattern – especially when the substances are highly penetrating – suggest differences of mutagenic action.

Since the rate of sperm utilization may be affected by the type of mutagenic treatment as well as by incidental conditions, the coordination between broods and treated germ cell stages requires the simultaneous observation of other genetical effects which serve as landmarks: thus crossing-over cannot be induced after prophase of meiosis, and bunches of identical mutations must stem from treated spermatogonia. The "landmark" technique has so far been applied only to X-rays and formaldehyde food; quite recently, Miss Sonbati in our Institute has used it for an analysis of the brood pattern produced by mustard gas. Both mustard gas and X-rays give a peak of mutation frequency at some stage between young spermatogonia and mature spermatozoa; but Miss Sonbati's results show that the sensitive stages to these two agents do not coincide. While X-rays act preferentially on late spermatids, mustard gas produces its strongest mutagenic effect on late spermatogonia or early spermatocytes. For reasons which have been mentioned before, it is not possible to decide whether this difference in brood pattern is due to an essential difference in mutagenic action or to secondary causes.

Of greater interest for the analysis of chemical mutagenesis is an experiment of Miss Sonbati's, in which she compared the ratio of translocations to lethals after mustard gas treatment of either mature sperm or germ cells in the sensitive stage. She found that this ratio changed in exactly the way predicted if during the sensitive stage the frequency of all genetical effects – mutation as well as chromosome breakage – had increased to the same degree. If, as has been claimed for X-rays, excess mutations during the sensitive stage were entirely due to intergenic changes, the frequency of translocations should have increased much more steeply than was actually observed. It is, of course, possible that this result cannot be applied to the interpretation of sensitivity to X-rays, especially since the sensitive stages to mustard gas and X-rays are not identical. But until compelling evidence to the contrary is obtained, it seems reasonable to assume that after irradiation as after mustard gas treatment the exceptional sensitivity during late or middle spermatogenesis is due to special physiological conditions that enhance the general mutagenic effectiveness of the treatment. Sobels and Oster have provided evidence that the main factor in X-ray sensitivity is degree of oxygenation. A similar analysis for mustard gas is still lacking, but preliminary tests by Sobels suggest that here, too, enzymes concerned with oxydative metabolism may be involved in the determination of sensitivity.

Two definite conclusions can be drawn from brood pattern analysis: (a) Mutations which take place in post-meiotic stages cannot occur through or in connection with crossing-over, and (b) postmeiotic mutations which give rise to non-mosaic mutants cannot be due to errors in gene replication. Many chemical mutagens produce lethals in post-meiotic stages and even in spermatozoa which are treated in the seminal receptacles of females. It is unlikely that all these lethals are mosaics in which the whole of the gonad is mutated; but real proof of non-mosaicism is provided only by the occurrence of visible mutations which affect the whole of the body, e. g. mutations to yellow body colour or singed bristles. By this test, only X-rays and mustard gas and, perhaps, some other alkylating agents have shown themselves capable of producing mutations by affecting the original treated gene rather than the process of gene replication. On the basis of the Watson-Crick model of DNA, even such mutations should give rise to mosaics, and the occurrence of non-mosaic mutants from treated spermatozoa would require special assumptions. It might, e. g., be assumed that in all cases of nonmosaicism the non-mutated strand of the DNA helix had been damaged so that it no longer took part in replication; but while this explanation may be acceptable for treatment with a drastic mutagen, it seems inadequate for spontaneous mutations. Yet, in an experiment which I carried out on *Neurospora*, non-mosaic mutations kept on occurring at a constant rate in stored dry conidia. These mutations seem to have arisen through some process that affected a whole cross-section of the DNA helix; it is conceivable that such a process is also responsible for the occurrence of non-mosaic mutants after chemical treatment of spermatozoa.

Mutagens which act exclusively by inducing errors in gene replication should always give rise to mosaic mutants. Conceivably, formaldehyde food acts in this way; for it acts only during the interphase preceding meiosis in male larvae, and mosaicism produced at this stage would no longer be discernible after meiosis. Another mutagen which might be suspected of acting mainly or only during replication is an alkylating agent, chloroethyl methane sulphonate. The very unusual sensitivity pattern of the *Drosophila* gonads to this substance was discovered by the Fahmys and has subsequently been studied in some detail by Miss Sonbati. Injection into adult males produces few lethals in postmeiotic stages; in later broods mutation frequency rises steeply and becomes very high in young spermatogonia. Injection into young adult females yields moderately high mutation frequencies, and these do not change very much in successive broods. Feeding the substance to larvae yields very high mutation frequencies in both males and females. If the few lethals which occur after post-meiotic treatment were mosaics, most of these peculiarities could be accounted for by the assumption that gene replication is a necessary condition for mutation. The results with adult females, however, throw doubt on this conclusion; for it is probable that gene replication had already taken place in the oocytes which were used for the production of the first brood. It thus seems more likely that chloroethyl methane sulphonate owes its peculiar brood pattern to the dependence of its mutagenic action on metabolic processes which take place in young germ cells of either sex, but have ceased to act – or act only weakly – after meiosis.

III. Mutagen specificity

The clearest evidence for mutagen specificity comes from experiments on micro-organisms; but some relevant results have also been obtained on animals, and I am going to discuss these briefly.

(a) The silkworm

Nakao has used the specific loci method for tests of mutagens. He compared the frequencies of two mutations, *pe* and *re*, which affect the colour of the embryonic membranes, so that large numbers of progeny can be scored by inspection of the eggs. The mutations are recessive and occur at loosely linked loci; *pe/pe* eggs are pink, and *re/re* ones are red. Nakao determined the ratio of *pe* to *re* mutations in the progeny of mutagenically treated wildtype males and untreated double-recessive females. In control matings, the ratio is 1 or more. After X-ray treatment, it is always higher than 1, and this was also so when treatment consisted in exposure to nitromine, the oxide of the nitrogen mustard HN-2, or to glycine "mustard". On the contrary, the *pe/re* ratio was 0.6 in two large experiments with alanine "mustard", and as low as 0.07 after treatment with the amide of alanine "mustard". The figures are large enough to exclude any doubt as to the reality of these differences between the effects of mutagens; but the interpretation is made doubtful by several curious features of this system. Apparent mutation frequencies were very high, even in the controls. Not many progeny tests were carried out, and the possibility has not been excluded that some of the red or pink phenotypes, which after chemical treatment usually occurred as mosaics, were due to temporary effects of the treatment on gene action, in the manner observed by Lewis for "revertible" changes of the self-sterility genes in higher plants. If the observed mutagen specificity should, indeed, be due to true genetic changes, it refers to chromosome regions rather than individual genes; for mutants that were progeny-tested usually turned out to carry deficiencies. The X-ray data, however, suggest that the seat of specificity resides at a level which is far removed from the primary effect of the mutagen on the genetic material; for while the *pe/re* ratio is always higher than 1 when males are irradiated, it is always lower than 1 when females are irradiated. It will be interesting to study whether changes in the *pe/re* ratio occur between successive stages of spermatogenesis. Thus Nakao's results, while giving clear proof of observable mutagen specificity, should also warn us against too facile interpretations of similar findings.

(b) *Drosophila*

SOBELS compared the relative frequencies with which X-rays on the one hand, mustard gas on the other, produced mutations at a number of loci on the X-chromosome; his data provided no evidence of striking differences between the effects of the two agents.

SLIZYNSKA found that rearrangements produced by formaldehyde food had a smaller proportion of breaks in the heterochromatin than rearrangements produced by X-rays. This is in contrast to results on plant chromosomes, which usually show a high proportion of heterochromatic breaks after chemical treatment. The discrepancy may, however, be spurious; for while in plants all break points are scorable, rearrangements with both breaks in the heterochromatin cannot be recognized in the salivary glands of *Drosophila*. On SLIZYNSKA's hypothesis that synchrony of replication promotes the simultaneous opening of latent breaks, such undetected rearrangements would be expected to be frequent after chemical treatment.

The FAHMYS studied the regional distribution of breaks and lethals induced by TEM. For breakpoints, the results are difficult to understand. While a first comparison with X-ray-induced breaks showed no significant difference ($P = .9$), subdivision of the X-chromosome into twice the number of segments resulted in a highly significant difference (P less than 10^{-11}). This discrepancy suggests the interesting possibility that TEM tends to break the X-chromosome in a few narrow regions, but no such analysis of the data seems to have been carried out.

The situation is clearer for the distribution of TEM-induced sex-linked lethals. The FAHMYS and BELITZ found independently that this distribution differs significantly from those of spontaneous and X-ray induced lethals. The difference resides mainly at the two ends of the X-chromosome. This suggests a possible centromere influence, and it would be interesting to follow up a suggestion made by Dr. MULLER and determine the distribution of lethals in an inverted chromosome. NAKAO's findings in the silk-worm suggest that regional mutagen specificities may also depend on the type of treated germ cell, and a comparison between the distributions of lethals in various stages of male and female germ cell development might be rewarding.

Visible mutations after treatment with TEM and other chemical mutagens had a similar distribution as lethals, so that some specificity on a regional basis might be expected. The FAHMYS' claims, however, go far beyond; they are rather reminiscent of GERSHENSON's claims for the specificity of thymonucleic acid, which were mentioned earlier in this lecture. The first claim is that certain chemical mutagens, e. g. TEM, phenylalanine "mustard", and chloroethylmethanesulphonate, produce a higher ratio of visible to lethal mutations than do X-rays. The obvious explanation, namely a relative shortage of small deficiencies among chemically induced mutations, does not hold; for the proportion of deficiencies among chemically induced lethals has so far always been at least as high as among X-ray induced ones. A distinction, thus, is made between lethal and visible point mutations. As I said before, I find it almost impossible to believe that this distinction is more than a convenient way of phenotypic classification, and that it should reflect an underlying distinction between "lethal" and "visible" genes. As a matter-of-fact, the evidence for this first claim of the FAHMYS is very slight. In experiments which were carried out independently by Mrs. ALTENBURG and myself neither phenylalanine "mustard" nor chloroethylmethanesulphonate produced more visible mutations than would have been expected in X-ray experiments yielding the same frequencies of lethals. It is evident that the frequency of detected visibles must vary between observers, and when the FAHMYS themselves scored visible mutations in the progeny of irradiated flies, they found a ratio of 2 visibles to 10 lethals, as compared with 3 visibles to 10 lethals after treatment with phenylalanine "mustard". In view of the inherent inaccuracy of figures for visible mutation frequencies, it seems unlikely that this difference represents a reality.

The second claim of the Fahmys is more interesting. After treatment with certain chemicals, in particular phenylalanine "mustard", they found repeatedly visible mutations which had not been described previously, and they assert that some alkylating agents produce mutations in genes which have not before mutated either spontaneously or after irradiation. Many of these new mutants are semi-lethal or sterile and for this reason may not have been included in published lists of *Drosophila* mutants. Others have very slight phenotypic effects, which may have escaped the notice of workers who were less intent on the discovery of visible mutants. But when these cases were deducted, there remained about a dozen good mutants which could be made into homozygous stocks. Females of these stocks are now being used for testing whether the new mutations can be produced by X-rays. The first results of these tests were reported by the Fahmys at the International Congress of Radiation Research in 1958. Several of the new mutations had been produced by treatment with phenylalanine „mustard", but not by irradiation. Although the data were not large enough for satisfactory tests of statistical significance, they certainly are suggestive. It would, indeed, be expected that the genes of *Drosophila*, like those of micro-organisms and bacteriophages, show some degree of mutagen specificity. It is, however, difficult to accept the claim that the new *Drosophila* mutants arise exclusively under the influence of a few chemicals, and the complementary claim that the specific action of these chemicals does not apply to known "conventional" mutants. On the contrary, the results of the Fahmys suggest the more moderate assumption that their chemicals produce a characteristic frequency distribution of mutations at different loci, and that the "new" mutants form the extreme end of this distribution. This assumption could be tested through allelism tests between independently produced autosomal lethals. This method has been used years ago by Hadorn on phenol induced lethals and did, in fact, suggest that phenol acts preferentially on certain loci on the second chromosome.

There is, however, one important difference between mutagen specificity in *Drosophila*, as suggested by the results of the Fahmys, and mutagen specificity in micro-organisms. In micro-organisms, specific differences are found between alleles at the same locus, or even between sites within the same cistron, whereas visible and lethal mutations in *Drosophila* are "forward" mutations which, presumably, include mutations at many sites within the same locus. So far, there have been only two indications of possible allele specificities of chemicals; in the Fahmys' experiments, the same new allele of *Notch* has occurred repeatedly, and in a small test with chloroethylmethanesulphonate two different males produced two phenotypically indistinguishable rare alleles of *white* which, according to tests kindly carried out for me by Dr. Green, had occurred at the same sublocus. For the rest of the new Fahmy mutants it has, on the contrary, been claimed that they are not allelic with any known mutant. This suggests that, after all, their specificity may be regional rather than genic.

Conclusion

In conclusion, I want to summarize what, to me, seems the value of mutation research on animals. There can be no doubt that, at the moment, micro-organisms down to the RNA of tobacco mosaic virus have not only provided the most exciting results in the field of chemical mutagenesis, but also hold out the greatest promise for spectacular progress in the near future. Yet I would warn against the attempt to concentrate all efforts into the nursing of this one growing-point. Science should not be like a straight-growing annual plant without side shoots, but rather like a shrub with many branches. Even branches which at the moment are overshadowed by others may in time bear fruit, and these will become important when, as is almost inevitable, the main shoot becomes less fertile. *Drosophila*, in particular, has still much to offer to students of chemical mutagenesis. For reasons which I gave at the beginning of my lecture, *Drosophila* lethals are the best tool for the detection of mutagenic ability in new compounds. For the same reason, the influence of factors such as oxygen supply or metabolic rate on mutation frequency is measured most conveniently in terms of sex-linked lethals. Once a new mutagen or a new factor in mutagenesis have been discovered in *Drosophila*, it will often be useful to turn them over to micro-organism geneticists for a more refined genetical and biochemical analysis, or to geneticists working on mice or mammalian cells for studies on possible applications to man or domestic animals. In the latter case, a preliminary brood pattern analysis on *Drosophila* may be valuable.

All these tasks require nothing but routine tests for sex-linked lethals. But *Drosophila* workers should not forget that their organism offers unrivalled opportunities for genetical and cytological studies by methods which can be adapted to the most varied purposes. To restrict investigations on *Drosophila* to tests for sex-linked lethals is like using a concert piano for playing nothing but scales. I wish that my lecture may have given the non-Drosophilists among you a slight idea of the variety of ways in which *Drosophila* can be used for attacking problems in chemical mutagenesis.

Chemical mutagenesis in higher plants

By Åke Gustafsson

Forest Research Institute of Sweden, Stockholm 51

Background. – In 1937 Stubbe summarized the data obtained up to that time concerning mutations induced by chemical means. He pointed out that Erwin Baur initiated such experiments already 1916 and stated the problem still pertinent in chemical mutagenesis: how to induce mutations of specific genes at will. In addition, he reported that the dominant mutation *Acorrugata* of *Antirrhinum* readily originates by chemical treatments. He paid much attention to the results of Soviet-Russian geneticists (Ssacharov and co-workers), who induced mutations in *Drosophila* by means of iodine treatments. These pioneer experiments are of interest in plant genetics, too, since iodine (and chlorine) acts as a mutagen also in barley (Ehrenberg *et al.* 1956; Wettstein *et al.* 1959) and breaks the chromosomes in *Oenothera* (Oehlkers 1947).

Scrutinizing these and other data, there seems to be no doubt that the problems of chemical mutagenesis were not only discussed long before 1940 (by Baur, Stubbe and others), but definite positive results had also been obtained (by Ssacharov and others). However, it was only in the early forties that chemical mutagenesis became a useful tool in mutation genetics. In 1943 Oehlkers described treatments with urethane (+KCl) causing chromosome breakage. He pursued this discovery in numerous papers, applying inorganic compounds, for instance iodine-potassium iodide and aluminium salts, organic compounds of various kinds, narcotics like ethyl urethane, acetophenon and glycol, as well as alcaloids like morphine and colchicine (Oehlkers 1947). In 1947 Auerbach *et al.* presented definite data on the mutagenic activity of mustard gas substances. The experiments started in 1940 and a report was submitted to the Ministry of Supply (Great Britain) in 1942. According to these experiments the sulphur as well as the nitrogen mustard exerted a mutagenic effect in *Drosophila* similar to that of X-rays. Continued studies revealed characteristics of these chemical mutagens differing in many respects from those of ionizing radiations. When applying mustard gas substances in barley (Gustafsson and Mac Key 1948; Mac Key 1954) it was found that they are mutagenic also in this species and that the spectrum of mutation is distinctly changed.

In the forties, too, Rapoport began his studies on the effects of epoxides and epimines in *Drosophila* and established their high mutagenicity (1948). Subsequently his results were confirmed for other organisms. These and other compounds were shown to influence backmutation rates of individual loci in *Neurospora* in quite a specific manner (Westergaard 1957). Through the work of Ehrenberg and his associates these compounds were soon found to be highly effective in phanerogams, in fact, as effective or more so than ionizing radiations.

Species and mutagens tested in phanerogams. – In the following only those publications will be considered, which have a direct bearing on the hereditary transmission of induced changes. Thus no or very few studies pertaining to the cytological effects of chemicals in mitosis (*Allium* or *Vicia faba*) will be mentioned. The hereditary changes are of two kinds: chromosome breaks leading to losses, gains or rearrangements of the chromosomes, and gene or point mutations where no concomitant gross changes in chromosome structure can be detected or deduced. Chemical compounds exerting cytogenetic effects have been divided into two groups, *mutagens* and *mutachromosomesos* (*cf.* FAVRET 1960b). Since there is no definite demarcation line possible between the two kinds of genetic changes, this distinction will not be used here, in accordance with the views expressed by OEHLKERS (1949) in his polemic with AUERBACH (1949). OEHLKERS claimed that the urethane-potassium chloride mixtures had proved "mutagenic" already 1943, though he only demonstrated the occurrence of chromosome translocations and not of visible or lethal mutations.

By and large the effect of chemicals with regard to plasmon or plastid mutations is unknown. MICHAELIS (1958) induced such mutation types with radioactive phosphorus and sulphur (P-32 and S-35) in *Epilobium* and WETTSTEIN with X-rays in *Hordeum* (unpubl.). FAVRET (unpubl.) obtained such a mutation type in *Hordeum*, too, after formaldehyde treatment.

The following species have been used for experimentation:

Antirrhinum majus – ethyl urethane (+ KCl), aluminium chloride, nitrogen mustard, codeine, thymonucleic acid (OEHLKERS 1956, BERGFELD 1958): numerous visible mutations of the lethal and viable type (*cf.* also the papers cited by STUBBE 1937).

Avena sativa – diepoxybutane (MURPHY and PATTERSON 1958): fatuoid and compactoid mutations.

Campanula persicifolia – ethyl urethane (+ KCl) (OEHLKERS 1943): chromosome translocations.

Hordeum vulgare convar. *distichon* – v. Table 1.

Lilium candidum – ethyl urethane (+ KCl), propyl urethane, aluminium chloride (OEHLKERS 1943): chromosome translocations.

Melilotus albus – ethyl urethane (+ KCl), aluminium chloride, glycol (+ KCl) (SCHEIBE and HÜLSMANN 1958): chlorophyll mutations, various defects, hereditary changes in coumarine content.

Oenothera sp. – numerous inorganic and organic compounds, among them aluminium salts, urethanes (+ KCl), alcaloids (OEHLKERS 1943, 1949, 1953; MARQUARDT 1949; LINNERT 1953; *cf.* also RÖBBELEN 1959): chromosome breakage and translocation.

Paeonia tenuifolia – ethyl urethane (+ KCl) (OEHLKERS and MARQUARDT 1950): half-chromatid, chromatid and chromosome translocations.

Pisum sativum – ethylene oxide, ethyleneimine, oxygen (under pressure) (BLIXT *et al.* 1958; GELIN *et al.* 1959): "visibles" of the lethal and viable types.

Sorghum sp. – colchicine (numerous papers, for references see FRANZKE and ROSS 1957; SANDERS *et al.* 1959): as the consequence of colchicine treatments mutants of various morphological appearance and immediately true-breeding are said to arise following upon mutation and somatic reduction division with a subsequent doubling of the chromosome number. The data are difficult to interpret according to traditional mutation theory. It should be noted that OEHLKERS (1949) found a high rate of chromosome translocations when applying colchicine to cut inflorescence of *Oenothera*.

Triticum aestivum – mustard gas and nitrogen mustard (MAC KEY 1954): speltoid and other morphological mutations.

Vicia faba – comparing X-rays and chemical mutagens like 8-ethoxycaffein and myleran, also progeny studies (J. and M. MOUTSCHEN-DAHMEN 1958 a, b and unpubl.; *cf.* also FORD 1949; REVELL 1953 *etc.*, concerning the localization of chromosome breaks).

Zea mays – mustard gas, ethylene oxide, diepoxybutane, urethane (+ KCl), aluminium chloride (GIBSON et al. 1950; FABERGÉ 1955; KREIZINGER 1958; RESPONDEK 1958): chromosome breakage (deficiencies), translocations, visible mutations.

Several types of plant material have been used when administering the chemical mutagens: (1) dormant or germinating seeds (most species and chemicals), (2) pollen grains (*Antirrhinum*: ethyl urethane (+ KCl), *Zea mays*: ethylene oxide, mustard gas), (3) cut stalks or inflorescences placed in solutions of the chemicals (this is the classical method adopted by OEHLKERS in *Oenothera*, successfully applied also in *Melilotus* and *Zea*), (4) young buds injected with chemicals (*Paeonia*). Generally the seed treatments seem to constitute the method most satisfactory and most easily reproduced.

Mutations and mutation rates. – The types of mutation generally studied concern (1) seedling characters, mostly chlorophyll deficiencies of the lethal or semilethal type, (2) various morphological and physiological properties, including fungus resistance and viable or even high-productive mutations, (3) sterility, either due to structural (translocation) heterozygosity or to recessive factor mutations that cause a break-down of meiosis and an elimination of gones or zygotes. – Chlorophyll mutations are useful when testing the suitable doses or concentrations of a mutagen. Viable mutations are specially useful for linkage studies and breeding purposes. Moreover, they provide valuable information about the mode of action of a mutagen, as well as the mutation pattern of a specific gene locus. Such information has been gathered for barley by studying the erectoid mutations (HAGBERG *et al.* 1958), mildew resistance (FAVRET 1960 a and b), as well as the bright-green (waxless) mutations (LUNDQVIST and WETTSTEIN unpubl.). With regard to ionizing radiations MAC KEY (1954) has gathered ample data on the speltoid locus in chromosome IX of hexaploid wheat. Finally, the translocations provide information concerning the chromosome breaking ability of a mutagen relative to its induction of visible mutations. In addition, the translocations may lead to new gene equilibria with subsequent small and often successful changes in physiology and productivity (NYBOM 1954, 1956; WETTSTEIN *et al.* 1959).

Most chemicals produce low rates of mutations, judging by the rates of chlorophyll mutations. A summary of the results found in barley is presented in Table 1. Considering this species, the outstanding chemicals that raise the mutation rates up to the level of ionizing radiations or higher comprise ethylene oxide and its derivative glycidol, ethyleneimine and some alkanesulphonic esters. The maximum rates of different mutation types using ethylene compounds were established by EHRENBERG *et al.* (1959) in greenhouse and field experiments (Table 2).

The maximum rates of chlorophyll mutations obtained by means of ethylene oxide almost equal those induced by X-rays and neutrons. In earlier experiments (v. WETTSTEIN *et al.* 1959) even slightly higher values are reached. The maximum rates of dwarf mutations are approximately equal. Viable mutations are decidedly more numerous after certain EO-treatments than those obtained after ionizing radiation. The maximum rates of "translocation sterility" amount to values around 40 per cent whether ethylene oxide, X-rays or neutrons have been used.

The frequency of chlorophyll mutations induced by means of ethyleneimine is three or four times higher than that obtained after ionizing radiations (first shown by EHRENBERG

Table 1

Maximal mutation rates obtained in barley with chemical compounds (chlorophyll mutations in the X_2-generation)

Chemical compound	Maximal mutation rate %	Author
I. Alkanesulphonic esters		
Ethyl methane-sulphonate	50.4	HESLOT *et al.* (1959)
,,	48.9	EHRENBERG *et al.* (unpubl.)
Propyl methane-sulphonate	25.4	,,
Methyl methane-sulphonate	11.1	,,
,,	7.8	HESLOT *et al.* (1959)
Chloroethyl methane-sulphonate	3.3	,,
Bromoethyl methane-sulphonate	3.1	,,
Di-methanesulphonyloxybutane (myleran) 1956	5.1	WETTSTEIN *et al.* (1959)
,,	1.6	HESLOT *et al.* (1959)
,, 1959	1.1	EHRENBERG *et al.* (unpubl.)
Di-methanesulphonyloxypentane	0.0	,,
II. Epoxides and epimines		
Ethyleneimine	33.0	EHRENBERG *et al.* (1958)
Glycidol	15.2	,, (unpubl.)
Ethylene oxide	9.4	WETTSTEIN *et al.* (1959)
Epichlorohydrin	2.9	EHRENBERG *et al.* (unpubl.)
Triethylenemelamin	1.5	,, (1960)
,,	0.5	HESLOT and FERRARY (1958)
Diepoxybutane	1.7	,,
,,	1.3	EHRENBERG and GUSTAFSSON (1957)
Ethylene glycol	0.9	EHRENBERG *et al.* (unpubl.)
III. Various mustard gas derivatives	1.7	HESLOT and FERRARY (1958); GUSTAFSSON and MAC KEY (1948); MAC KEY (1954); EHRENBERG *et al.* (unpubl.)
IV. Purine and purine derivatives		
8-ethoxycaffein (chromosome breaking)	0.6	EHRENBERG *et al.* (unpubl.); cf. EHRENBERG *et al.* (1956)
Nebularine	1.6	WETTSTEIN *et al.* (1959)
2-aminopurine	0.6	EHRENBERG *et al.* (unpubl.)
Adenosin	0.6	,,
V. Various alkylating and oxidizing compounds, including a great number of organic and inorganic chemicals (also oxygen):		
Diethylsulphate	10.0	HESLOT et FERRARY (1958)
,,	10.2	EHRENBERG *et al.* (unpubl.)
2-bromoethylamine/hydrobromide	5.6	HESLOT *et al.* (1959)
Oxygen	a few per cent	MOUTSCHEN *et al.* (1959); KRONSTAD *et al.* (1959); EHRENBERG (1960)
Formaldehyde	ca. 0.5	FAVRET and RODRIGUEZ (1957)
Numerous mutagenic agents	slightly raised mutation rates	EHRENBERG *et al.* (unpubl.); D'AMATO (1950)
VI. Controls (different varieties)	0.05–0.1	EHRENBERG *et al.* (1956, 1959 and unpubl.); HESLOT *et al.* (1959); FAVRET (1960)

Table 2

Rates of different mutation types obtained with ethylene oxide, ethyleneimine and ionizing radiations in parallel experiments

Treatment	Chlorophyll mutations[1]) %	Dwarf mutations[2]) %	Viable mutations[2]) %	"Translocation sterility"[2]) %
X-rays				
16000 r	(9.1)	2.0	2.0	20.0
12000 r	5.3	(–)	(–)	(41.7)
8000 r	6.8	2.7	5.4	38.9
Neutrons				
640 rep	10.0	2.5	1.7	25.2
400 „	6.3	3.4	3.6	38.4
Ethylene oxide				
0.15% EO-solution, dry seeds	6.0	1.0	2.2	10.7
0.30% „ „	5.8	4.2	7.7	40.6
0.20% „ „	5.4	2.2	8.9	27.0
0.25% „ , soaked seeds	5.9	(–)	(8.0)	(–)
Ethyleneimine				
0.04% EI-solution, dry seeds	27.0	6.2	12.7	38.7
0.07% „ „	23.4	7.1	11.9	40.3
0.05% „ , soaked seeds	17.7	6.5	12.4	34.7
0.10% „ „	12.9	4.3	16.5	36.5
0.15% „ „	11.8	6.0	9.1	21.2
0.20% „ „	9.8	13.2	18.3	39.5

[1]) Greenhouse tests.
[2]) Field experiments.
Within parenthesis offspring consisting of less than 30 spike progenies.

et al. 1958). Dwarf mutations are at least two or three times more frequent. Viable mutations, relatively seen, increase as much as chlorophyll mutations or more. The frequency of "translocation sterility", on the contrary, does not, even with ethyleneimine, surpass the maximum rate of 40 per cent, characteristic of ionizing radiations.

Data concerning the effects of the diethylsulphate and ethyl methanesulphonate (HESLOT and FERRARY 1958; EHRENBERG *et al.* unpubl.) are less complete. Only rates concerning chlorophyll mutations have been published as yet. They run as follows (HESLOT and FERRARY *l. c.*; HESLOT *et al.* 1959):

	No. of spike progenies	Chlorophyll mutations %
Control	10000	0.05
γ, 25000 r	1000	4.5
diethylsulphate		
0.2%	732	9.8
0.1%	100	10.0
ethyl methanesulphonate		
1 %	260	50.4
0.4 %	210	43.3
0.25%	500	31.0

Diethylsulphate causes mutation rates twice as high as those obtained with γ-rays. Swedish experiments in 1958/59 produced rates about 10 per cent. The ethyl methanesulphonate is extremely efficient, in the experiments of HESLOT *et al.* (1959) it was ten times as mutagenic as the most efficient dose of γ-radiation. This result has been corroborated and further extended by EHRENBERG *et al.*, who applied a series of related compounds.

Compared with the chemicals discussed above other compounds analysed in barley are considerably less efficient. In *Pisum*, too, the ethyleneimine is highly mutagenic. Data are not yet available for other phanerogam genera, although work is under way. With regard to other chemicals, like the urethanes, only a few experiments with parallel chemical treatments and irradiation have been performed. OEHLKERS found in 1949 (p. 28) that the effect of 1/20 mol. ethylurethane and 1/200 mol. KCl approximately corresponded to a dose of 200–300 röntgens of X-rays. According to OEHLKERS (1956) aluminium chloride and ethyl urethane (+ KCl) in *Antirrhinum* certainly increased the rate of visible mutants in the second generation but less so than nitrogen mustard. Comparing the mutation rates after chemical treatments of cut inflorescences with those obtained in STUBBE's experiments (1932) after an X-irradiation of buds, it is found that the frequencies induced by X-rays attain decidedly higher values. In three sets of experiments, carried out by BERGFELD (1958), the following mutation frequencies were recorded:

(1) aluminium chloride → codeine → thymonucleic acid → distilled water → morphine → control,

(2) ethyl urethane (+ KCl) → aluminium chloride → codeine,

(3) aluminium chloride → codeine → thymonucleic acid → distilled water.

No comparison with X-irradiation was made. When a method of treatment similar to that used in *Oenothera* and *Antirrhinum*, was applied in *Melilotus*, ethyl urethane (+ KCl) was highly efficient. Mutation rates of chlorophyll deficiency amounting to circa 20 per cent were obtained. Other chemicals produced lower rates. With regard to mutants of the viable type, containing less coumarine than the controls, aluminium chloride seemed to be more effective than the urethanes; not less than eight out of twelve such mutants were produced in series where this chemical was used. Three were obtained when glycol (+ KCl) was used and one after a treatment with ethyl urethane (+ KCl). No comparable data of X-irradiation have been presented.

With regard to treatments with mustard gas derivatives, GIBSON *et al.* (1950) pointed out for maize that mustard gas caused chromosome breakage but yielded deficiencies instead of translocations. The amount of aborted pollen in the various treatments corresponded to that obtained after an X-ray dosage of circa 250r. Point mutations seem to be rare. Losses of dominant markers are more obvious in the endosperm than in the embryos, where the material treated shows less losses than the controls. In barley the mutation rates induced by a nitrogen mustard treatment correspond to those induced by 1000r of X-rays or less (MAC KEY 1954). In *Antirrhinum* OEHLKERS (1956) found nitrogen mustard to be the most efficient chemical applied.

Summarizing the results obtained, we may state that several chemicals are found to raise the mutation rates in higher plants greatly, some to considerably higher levels than those reached by ionizing radiations. However, the possibilities of inducing high mutation rates by means of chemicals are certainly not exhausted by the detection of the ethyleneimines and alkanesulphonic esters hitherto studied (EHRENBERG *et al.* 1959, p. 367).

Mutation spectra and controlled mutation. – Almost twenty years ago it was shown for *Hordeum* and *Antirrhinum* that there exists a sort of group mutability in these genera, i. e. special types of mutation arise after certain modifications of the X-ray treatment. This group mutability seemed to indicate a characteristic basic constitution of the numerous genes or chromosome parts (segments) that are responsible for the individual types of mutation

(D'AMATO and GUSTAFSSON 1948). KAPLAN (1946) suggested that a number of genes may possess some distinct fine-structure in common (e. g., dissociation constants). The isophenic segments occurring in barley (FAVRET 1960; HAGBERG 1960) favour such an idea. Among the various chlorophyll types, the *alboxantha* phenotype, which rarely appears and possibly is caused by mutation in one or a few genes, seems to arise predominantly after the irradiation of pregerminated seeds (at least in the varieties Golden and Maja barley, GUSTAFSSON 1947). This indicates that the cell nuclei must be in a certain physiological state for *alboxantha* mutations to arise. If the high hydration is not present, even the highest X-ray doses seem unable to induce this type of mutation (*l. c.*, p. 58).

In 1948 and 1949 D'AMATO and GUSTAFSSON, as well as GUSTAFSSON and NYBOM, showed that a colchicine treatment previous to irradiation led to a distinct change in the mutation spectrum: rare chlorophyll types were formed more frequently after such a pretreatment. In addition, EHRENBERG *et al.* (1956) stressed the selective elimination of certain types of chlorophyll mutations in experiments with high dosages of irradiation.

With the widened use of chemical mutagens in genetical and cytological research, in *Drosophila* and in *Allium* and *Vicia faba* roots, it soon became evident that there are differences in action between chemical mutagens and ionizing radiations. The OEHLKERS group was able to show that the chromosome breaks induced in *Oenothera* and *Vicia faba* are not so entirely random, when metal salts, urethanes or nucleic acids are applied, as they are after X-irradiation (see OEHLKERS 1953 for references). The "weak spots" consist of achromatic regions (centromeres and satellite zones) similar to the case in spontaneous breakage. Studying the breaks, produced by γ-rays and nitrogen mustard in the long chromosome and in the five short ones, FORD was the first research worker to show such an effect in roots of *Vicia faba* (1949). He found the proportion for breaks appearing in the long chromosome and the short chromosomes to be 5:2 when applying γ-rays and a ratio of 50:2 when using nitrogen mustard. REVELL (1953) further elaborated on this point. He concluded that the breaks induced by NN-di(2-chloroethyl)methylamine, a certain diepoxide, di-(2:3-epoxypropyl)ether, and the tertiary butyl hydroperoxide are clustered to the heterochromatic zone of the long chromosome. According to DARLINGTON and MCLEISH (1951) maleic hydrazide induces breaks in the heterochromatin only. Using purine derivatives KIHLMAN and LEVAN (1951) similarly found a strict localization of the breaks in *Vicia faba* to the heterochromatic regions. The effect of 8-ethoxycaffein is remarkable. This compound induces a good deal of chromosome breakage also in barley (J. and M. MOUTSCHEN-DAHMEN 1958a) but it does not produce a distinct rise in the rate of chlorophyll mutations. Breaks, produced by 8-ethoxycaffein, are concentrated to the centromeres and to the secondary and tertiary constrictions. The single chemical so far known to induce random chromosome breakage, like the case with X-irradiation, is oxygen applied at high pressure (MOUTSCHEN-DAHMEN *et al.* 1959). High mutation rates are obtained in barley (cf. also KRONSTAD *et al.* 1959), as well as in *Pisum* (GELIN *et al.* 1959).

All these studies, and others as well, indicate that ionizing radiations and chemical mutagens affect the chromosome material in different manners, as judged from chromosome breakage. A postulated difference in action formed the background to the work on chemical mutagenesis in crop plants initiated at Svalöf as early as the forurties. It was found (GUSTAFSSON and MAC KEY 1948) that the chlorophyll mutations obtained essentially were of the viridis-type when applying sulphur and nitrogen mustard, whereas albinas were rare compared to the case with X- or neutron irradiation. This change in mutation spectrum has then subsequently been confirmed by MAC KEY (1954).

In a recent paper WETTSTEIN *et al.* (1959) have given a full account of the changes in mutation spectrum that occur when different mutagens are used. Albina mutants become progressively rare, and viridis mutations common in a series of treatments ranging from neutrons and X-rays to ethylene oxide and ethyleneimine and, finally, to di(β-chloroethyl)phenylalanine and nebularine (purine). The spectrum of spontaneous mutations is situated somewhere in the middle (between those of myleran and nitrogen mustard). As previously mentioned, the frequency of various types of mutations (excluding chromosome translocations) is 3–4 times higher after ethyleneimine treatment than that obtained with X-rays and neutrons. This relationship holds true for chlorophyll mutations, dwarfs and viable mutations as well. The rise in mutation rates could possibly be due to a general increase in gene mutability, spread in a like manner over the genome. This presumption is not valid, however, since the spectrum of chlorophyll mutations varies with the type of mutagen. Hence, a parallel rise in mutation rate certainly exists but it conceals differences in the mutagenic action.

This is further evidenced by the relative frequencies of "translocation sterility" and visible mutants. No less than 60–70 per cent of the aberrant X- and neutron progenies are characterized by such sterility, and more so with neutrons than with X-rays. The corresponding figure for ethylene oxide and ethyleneimine lies between 45 and 48 per cent. Consequently, more visible mutations (point mutations) and less chromosome aberrations are induced by these two chemicals, relatively seen, than by the ionizing radiations.

Chromosome breakage and gene mutations, the two basic events in mutation, are in a sense independent of one another (EHRENBERG *et al.* 1956; *cf.* also GAUL 1958). This implies that a chromosome break – with or without a subsequent structural rearrangement – may not necessarily induce a gene change on the molecular level, although the rearrangement may sometimes lead to an "expression effect" (STADLER 1954). In fact, many extreme alterations of the barley karyotype, e. g. those involving chromosomes 6 and 7, have no distinct effects on morphology and productivity. On the other hand, breaks in special segments of chromosomes 1 and 5 are often combined with erectoid mutation and changes in mildew reaction respectively.

Likewise, a gene mutation may arise independently of chromosome breakage. Mutagens are known which do not lead to chromosome breakage (as judged from the immediate cytological effects and from the X_1 or X_2 sterility) but cause a rise in the mutation frequency. Consequently, gene mutation may take place without any intervention of chromosome breakage and, conversely, random chromosome breakage will not influence gene structure and lead to gene mutation. This still recognizes the fact that in many cases chromosome breakage and rearrangement will force neighbouring genes to mutate, depending on the gene loci in question and the place of breakage.

Hence, by the application of the right mutagens or by a special pre-or post-treatment, physical or chemical, we may differentiate between the basic processes of mutation, although no sharp delimitation line between them is possible.

This is not the place to discuss the mechanism of chromosome breakage itself. The analysis carried out by WOLFF (1957) led to the view that ionizing radiations may produce two sorts of chromosome breaks, such remaining open for a short time, others requiring energy for their healing and staying open considerably longer. The short-lasting breaks can be prevented from restitution by the addition of chelate-forming compounds. In this way, the chromosome rearrangements that arise through recombination of breaks, can be influenced in both ways: they may increase or decrease in number.

J. and M. MOUTSCHEN-DAHMEN made a comparative study of the effects of 8-ethoxycaffein and myleran (1958 a and b). In *Hordeum* both mutagens produced localized breaks: to the centromere regions, as well as to the secondary and tertiary constrictions. There is one remarkable difference between these mutagens, however. Translocations are not formed when myleran is used and the total chromosome length seems to diminish in consequence of the treatments. Combined with X-irradiation, myleran increases the amount of chromosome disturbances produced by irradiation alone. This increase is restricted to chromosome breakage, however: the formation of translocations ("bridges") is greatly inhibited and distinctly below the level reached after X-irradiation alone (unpubl.).

Hence, by the use of different chemicals, or by a combination of ionizing radiations and chemicals, we may alter not only the mutual proportion of point mutations and chromosomal rearrangements, but also the types of structural changes appearing. Consequently, the mutation spectrum can be considerably modified by applying different kinds of mutagens. However, numerous genes are in many cases responsible for the appearance of a special phenotype, say the albina-type of chlorophyll mutation. The change in mutation spectrum, therefore, depends on a sort of group mutability, as mentioned previously (p. 19). Only in those cases when a special phenotype can be attributed to mutation in a single gene, the possibility exists to determine the mutagenic reaction of the gene loci by a mass-analysis of the phenotypes. Eventually, for a full analysis of the mutation process, we must proceed to a study of specific genes and their reaction to different mutagens. There are but a few such instances known in higher plants. One of them concerns the erectoid loci (Hagberg *et al.* 1958; Hagberg 1960), one the mildew resistance (Favret 1960, a, b), both cases in barley.

The origin of erectoid mutations. – At least 22 different gene loci (a, b, c . . . . v) determine the origin of the erectoid character. Yield tests have been carried out with great many erectoid mutants. High-productive types have been found among mutants of loci a, c, d, l, m, o. One mutant (of locus m) has been approved as a new trade variety. Many erectoid loci form "isophenic segments". Such is the instance with regard to locus d, close to the end of the long arm of chromosome 1, and linked with a series of other *ert*-loci. Possibly there are one or two more isophenic segments in the barley karyotype influencing ear density.

Erectoid mutations are continually being produced by a series of mutagens, physical as well as chemical. The physical mutagens consist of γ-rays (acute or chronic), X-rays, neutrons, fast protons, α-rays. Recent work with ionizing radiations has been concentrated on the first three types (γ-, X- and neutron rays). Distinct regularities of mutation were found with regard to the frequently mutating loci a, c, d, m. In addition, loci b and r display special features as discussed in the papers cited above – locus b mutating in a special variety, not in others, locus r giving dominant mutations. Locus a, with 11–12 mutations, has so far not mutated after neutron treatments but frequently after γ- and X-irradiation. Chromosome breakage has not been found in any single case. In locus c all types of radiation, sparsely as well as densely ionizing ones, are effective. After the begin of neutron treatments (in 1947) this locus became the most frequently mutating one. Seventeen cases of mutation have been localized so far. Translocations in conjunction with these c-mutations are common; six have been isolated, three with γ- and X-rays, three with protons and neutrons. Such translocations involve chromosome 3, where locus c is situated; in three cases the second break is in chromosome 2. Locus d, with ten mutations, has mutated after γ-, X- and neutron treatments. In one case – *ert*-d^7 – a translocation is present. In locus m, with seven to eight mutations, three were induced by X-rays, four (or five) by neutrons. No translocations occurred in conjunction with mutations of this locus. The following distribution is obtained:

	γ-rays	X-rays	Protons Neutrons	With translocations	Sum of mutations
Locus a	1	10 (11)	–	–	11 (12)
,, c	2	6	9	6	17
,, d	1	6	3	1	10
,, m	–	3	4 (5)	–	7 (8)

The conclusion seems evident: Specific gene loci react in a distinct manner to different types of ionizing radiation, depending on their physical characteristics. GILES, KØLMARK and WESTERGAARD (*cf.* the last-mentioned author 1957) were able to show that the back-mutation of individual gene loci in *Neurospora* proceeds in a definite manner depending on the mutagenic agent applied. With regard to the erectoid mutations in barley apparently forward-mutation, too, can be regulated by chosing the right physical mutagens. Since chromosome breakage is often involved in mutations of locus c and since densely ionizing agents are ineffective in the case of locus a (or they destroy the cell when forcing the locus to mutate), individual chemical mutagens will probably differ also with regard to the origin of erectoid mutation, depending on the chemical structure and the mode of action of the chemical (*cf.* the effects of ethylene oxide and diepoxybutane, or nebularine and 8-ethoxycaffein). Although a great number of chemically induced erectoid mutations have been isolated and crossed, no definite data have been obtained in this respect so far. However, the problem is of great importance, since ultimately we may be able to control the origin of high-productive mutations by the use of the right chemical even better than with ionizing radiations.

According to unpublished data gathered by LUNDQVIST and von WETTSTEIN the bright-green (waxless) mutations in barley are produced by changes in a great many gene loci (possibly more so than in the case of the erectoids) and the bright-green loci react, like the erectoid loci discussed above, in a different manner to sparsely and densely ionizing mutagens.

Mildew resistance in barley. – FAVRET (1960a, b) has recently summarized our knowledge about mildew resistance and its changes by mutation. Eighteen factors for resistance distributed over 14 loci are known. No less than 17 factors lie in chromosome 5, the single remaining factor is referred to chromosome 4. The linkage relationships of the isophenic segment, 40–50 units long, of chromosome 1 are as follows:

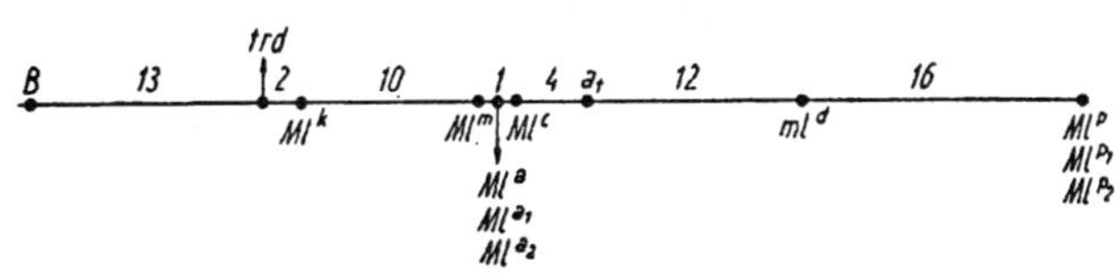

Abb. 1.

Locus Ml^p has at least three or four alleles for resistance. Three other loci, Ml^m, Ml^a and Ml^c, form a genic complex, involving six or seven different alleles.

Chromosome breakage within the isophenic segment of chromosome 5 often leads to changes in mildew reaction. This is the case with erectoid 7 – *ert*-d^7 – with two isophenic segments involved, one of erectoid loci in chromosome 1 and one of mildew loci in chromosome 5. Other similar translocations are under study.

Applying four mutagens – γ-rays, neutrons, formaldehyde and ethylene oxide – FAVRET studied the rates of chlorophyll mutations and mutations for resistance or susceptibility to mildew.

The data on mutability are summarized in the following table:

1	2	3	4	5
Treatment	No. of spike progenies	Mildew mutations	Chlorophyll mutations	Ratio (3:4)
(A) Susceptibility → resistance				
Spontaneous	38141	0	22	0.00
γ-rays	8531	14	213	0.07
Ethylene oxide	9251	0	41	0.00
Formaldehyde	15849	1	44	0.02

1	2	3	4	5
Treatment	No. of spike progenies	Mildew mutations	Chlorophyll mutations	Ratio (3:4)
(B) Resistance → susceptibility				
Spontaneous	23377	3	13	0.23
γ-rays	3783	1	147	0.01
Ethylene oxide	5612	2	57	0.04

Mutations for susceptibility obtained with the chemicals used are ten times as common as mutations for resistance (0.04 against 0.004%). A reverse tendency prevails after γ-irradiation (0.03 against 0.16%). For the two groups of mutagens, the ratio between mildew and chlorophyll mutations, too, is different in the two directions. In the case of resistant mutants (A in the table) the ratio is 0.07 for γ-rays and 0.01 for chemicals. In the case of susceptible mutants (B in the table) the ratio is 0.01 and 0.04 for γ-rays and chemicals respectively. γ-rays induce resistance more readily than they induce susceptibility, both in an absolute and in a relative sense.

Formaldehyde is a fairly weak mutagen as compared with ethylene oxide. This is visualized in the table by the number of chlorophyll mutations obtained. Still these two mutagens work in a similar manner and they distinctly differ from γ-irradiation. Chemical mutagens that primarily break the chromosomes, e. g. 8-ethoxycaffein and myleran, are expected to influence mildew mutation in a manner deviating from that of formaldehyde and ethylene oxide.

FAVRET also pointed out that mildew and chlorophyll mutations show a different dose relationship. The increase is exponential for mildew and linear for chlorophyll mutations (except at high dosages where chlorophyll mutations are eliminated).

1	2	3	4	5
γ-dose (r)	No. of spike progenies	Mutations for resistance (per spike progeny and r-unit)	Chlorophyll mutations (per spike progeny and r-unit)	Ratio (3:4)
3000	713	$0.0 \cdot 10^{-8}$	$233.3 \cdot 10^{-8}$	0.00
10000	3649	8.2	227.0	0.03
20000	651	23.0	230.0	0.10
30000	551	48.3	199.0	0.24

No (or few) mutations for resistance occur spontaneously. In the opposite direction, however, – for susceptibility – mutations seem to be fairly common. FAVRET concludes that a high level of resistance is primarily produced by mutagens acting like the ionizing radiations, i. e. profoundly breaking the chromosomes. Spontaneously, or with mutagens less effective in chromosome breakage, there are more chances of getting susceptible mutants than resistant ones.

Anyhow, largely concentrated to one segment of chromosome 5, the gene complex of mildew reaction seems to react to ionizing radiations in a manner differing from that of the chemical mutagens tested. Like in the case of chlorophyll mutations, ionizing radiations apparently give another spectrum of mildew mutations than that produced spontaneously.

The mode of action of chemical mutagens. – This problem was analysed by EHRENBERG (1959), comparing ionizing radiations and chemical mutagens. In a recent paper, to which the reader is referred, he specially deals with the chemical aspects of the problems (1960).

With reference to higher plants, the action of ethylene oxide and diepoxybutane is of special interest (EHRENBERG and GUSTAFSSON 1957). The investigation cited disclosed fundamental differences in the effects of the two compounds. Diepoxybutane was much more effective than ethylene oxide in causing lethality and sterility (about 300 times), probably depending on the double epoxide character. In addition, it certainly was the most active mutagenic agent of the two, although superior only by a factor of 50. Owing to the high toxicity of diepoxybutane, the mutation rates obtained with this chemical did not reach the values obtained in the ethylene oxide treatments. Apparently, alkylations on two adjacent points are expedient for the origin of chromosome rearrangements (sterility); just one single alkylation is sufficient for the induction of gene mutation.

This and other findings contrast to some ideas presented by GOLDACRE *et al.* (1949), LOVELESS and REVELL (1949) and LOVELESS (1951), *viz.* that the polyfunctional epoxides, epimines and other mutagens are more active than the simple ones in mutagenesis and carcinogenesis. As reviewed by WESTERGAARD (1957), the back-mutation rates of the adenineless gene locus in *Neurospora* were established for several chemical mutagens. The following rates were found (calculated on the number of mutations per conidium):

	Simple compounds			Polyfunctional compounds	
I A	Epichlorohydrin	$-56 \cdot 10^{-6}$	I B	Diepoxybutane	$-85 \cdot 10^{-6}$
	Glycidol	− 34		Diepoxypropylether	− 1
	Propylenoxide	− 21			
	Ethylene oxide	− 17			
	1,2-monoepoxybutane	− 3			
	Epibromohydrine	− 3			
II A	Ethyleneimine	− 16	II B	Triethylenemelamin	− 0.6
III A	Chloroethyl methane-sulphonate	− 51	III B	Di-methanesulphonyl-oxybutane (myleran)	− ?
	Ethyl methane-sulphonate	− 17			

With regard to this special gene locus and its back-mutation to the +-allele diepoxybutane was the most active mutagen of all those tested. In group II the polyfunctional triethylenemelamin was less efficient than the simple ethyleneimine. For the third group of alkanesulphonate esters no comparison was made with the complex compounds, e. g. myleran. Following the results obtained with the simple and double epoxide in barley it was suggested by EHRENBERG in 1956/57 that the polyfunctional compounds would be less efficient in higher organisms, like *Hordeum* and *Pisum*, than the simple ones and that the simple compounds should be applied in order to raise the mutation rates profoundly. Since myleran was fairly active in barley, also with regard to chlorophyll mutation, he proposed the use of its molecular half, the ethyl methane-sulphonate also in higher plants. Later on the superiority of the simple compounds was verified, with ethyl methane-sulphonate and ethyleneimine as the strongest mutagenic agents found so far in phanerogams (EHRENBERG *et al.* 1958 1959; HESLOT *et al.* 1959; EHRENBERG 1960, and unpubl.).

Summary and conclusions

In recent years extremely efficient chemical mutagens have been detected. Even in phanerogam species they induce mutation rates higher than those ever obtained with ionizing radiations. Two such compounds are ethyl methane-sulphonate and ethyleneimine. Rates of chlorophyll mutation five or six times as high as those found in irradiation experiments are reached. Of these mutagens ethyleneimine induces viable mutations in barley three to four times as frequent as X-rays and five times as frequent as neutrons. No data with regard to the ethyl methane-sulphonate are available as yet. Hence, by the use of certain chemical mutagens we may force the barley genotype to mass-mutate, and so, in principle, the genotype of any phanerogam species. Certainly, the rates attained are not the highest mutation rates possible.

Chemical mutagens often work in a manner different from that of ionizing radiations. Some, e. g. 8-ethoxycaffein, break the chromosomes without noticeably increasing the rate of visible mutations. Others, like myleran, break the chromosomes and cause visible mutations but prevent chromosome rearrangements, alone or in combination with ionizing radiation. Some, like nebularine, raise the rates of visible mutation without breaking the chromosomes.

Finally, certain compounds like the simple epoxides and epimines induce chromosome translocations and point mutations in a proportion different from that obtained by means of ionizing radiations. Actually chromosome aberrations are less frequent in progenies segregating for visible mutations when using such chemicals than with X-rays and neutrons.

The spectrum of visible mutations distinctly changes with the type of mutagen. This has been confirmed for the lethal or semilethal chlorophyll mutations. No doubt, it holds true for viable mutations as well. One case is known – with regard to mildew mutation – where a chemical mutagen, ethylene oxide, has another effect on forward- and back-mutation than γ – (and neutron) irradiation.

The erectoid mutations occuring in barley indicate that individual gene loci react differently to various types of ionizing radiation, depending on the type of linear energy transfer (ionization density). Locus a-erectoids, for instance, do not, or but rarely, arise after neutron treatments. Chromosomal rearrangements are unknown in conjunction with a-erectoids. Locus c-erectoids, on the other hand, are especially frequent after neutron treatments and, in addition, they often arise associated with chromosome breakage and translocation. Locus d- and m-erectoids appear just as well with sparsely as with densely ionizing agents independent of any detectable chromosome breakage. Most probably chemical mutagens will influence the mutation of individual erectoid loci in a definite manner.

Hence, we are no doubt entitled to state that chemical mutagenesis is a reality in phanerogam species, as well as in lower animals, fungi and bacteria. In fact, mutation work in higher plants cannot progress without the use of chemical mutagens. This does not imply, however, that the ionizing radiations have become obsolete. In the author's opinion, various kinds of mutagens, physical as well as chemical, should be subject to parallel applications until we know in more detail the best way of inducing chromosome breakage in specific regions and point mutation in specific loci. The erectoid and mildew mutations show how diversified are the reactions of individual segments and loci, also when using such clumsy tools as the ionizing radiations.

The origin and properties of erectoid mutations in barley have been considered in a long series of papers. Numerous data concerning their yield properties are available. Suffice it to state, that it is possible in two-row barley to induce erectoid mutations which are high-yielding, non-lodging and with good grain quality. In fact, we have begun to understand the methods of producing such mutations. Furthermore, as indicated by erectoid-d^7, it has been possible by means of chromosome breakage and translocation to induce a productive erectoid mutation in chromosome 1 and simultaneously a partially mildew-resistant mutation in chromosome 5. Here, in one stroke, we have gained a variety with a new karyotype possessing several outstanding characteristics of agricultural significance.

Finally, recent work by FAVRET, HAGBERG and others has thrown new light on the possibilities of inducing mildew resistance by artificial mutation. Numerous genes and alleles are involved in these reactions, most of them concentrated to one chromosome (No. 5) and one segment within this chromosome. Distinct regularities of mutation occur. These findings are the best rejoinder to the sceptics who question the possibility of an induced disease resistance.

The similarity in the induction of erectoid and mildew mutations suggests a general applicability of mutagens in the planned improvement of crop plants. As a method in plant breeding artificial mutagenesis cannot replace hybridization and recombination. Rather, it is an additional tool. Its significance will no doubt increase, however, to the extent we learn to predict the chromosome and gene changes which are produced by various mutagens.

Literature cited

AUERBACH, C., 1949: Chemical mutagenesis. – Biological Rev. 24, 355–391.

AUERBACH, C., J. M. ROBSON, and J. G. CARR, 1947: The chemical production of mutations. – Science 105, 243–247.

BERGFELD, R., 1958: Mutationsauslösung durch Chemikalien bei *Antirrhinum majus* L. – Zeitschr. f. indukt. Abstammungs- und Vererbungslehre 89, 131–142.

BLIXT, S., L. EHRENBERG, and O. GELIN, 1958: Quantitative studies of induced mutations in peas. I. Methodological investigations. – Agr. Hort. Genetica 16, 238–250.

D'AMATO, F., 1950: Mutazioni clorofilliane nell'orzo indotte da derivati acridinici. – Caryologia 3, 211–220.

D'AMATO, F., and Å. GUSTAFSSON, 1948: Studies on the experimental control of the mutation process. – Hereditas 34, 181–182.

DARLINGTON, C. D., and J. MCLEISH, 1951: Action of maleic hydrazide on the cell. – Nature 167, 407.

EHRENBERG, L., 1959: Radiobiological mechanisms of genetic effects. A review of some current lines of research. – Rad. Res. Suppl. 1, 102–103.

— 1960. Chemical mutagenesis: Biochemical and chemical points of view on mechanisms of action. – In press.

EHRENBERG, L., and Å. GUSTAFSSON, 1957: On the mutagenic action of ethylene oxide and diepoxybutane in barley. – Hereditas 43, 595–602.

EHRENBERG, L., Å. GUSTAFSSON, and U. LUNDQVIST, 1956: Chemically induced mutation and sterility in barley. – Acta Chem. Scand. 10, 492–494.

— 1959: The mutagenic effects of ionizing radiations and reactive ethylene derivatives in barley. – Hereditas 45, 351–368.

EHRENBERG, L., Å. GUSTAFSSON, and D. VON WETTSTEIN, 1956: Studies on the mutation process in plants – regularities and intentional control. – Conference on chromosomes, Wageningen, April 1956. Publ. by W. E. J. Tjeenk Willink, Zwolle, Netherlands.

EHRENBERG, L., U. LUNDQVIST, and G. STRÖM, 1958: The mutagenic action of ethyleneimine in barley. – Hereditas 44, 330–336.

FABERGÉ, A. C., 1955: Types of chromosome aberrations induced by ethylene oxide in maize. – Genetics 40, 571.

FAVRET, E. A., 1960a: Spontaneous and induced mutations of barley for the reaction to mildew. – Hereditas 46, In press.

— 1960b: Induced mutations for resistance to disease. – In press.

FAVRET, E. A., and A. A. RODRIGUEZ, 1957: Inducción de mutaciones en cebada con tratamientos de formaldehida. – Revista de Investigaciones Agricolas 11. No. 3, 305–307.

FORD, C. E., 1949: Chromosome breakage in nitrogen mustard treated *Vicia faba* root tip cells. – Proc. of the Eighth Int. Congr. of Gen., Suppl. Vol. Hereditas p. 570–571.

FRANZKE, C. J., and J. G. ROSS, 1957: A linear series of mutants induced by colchicine treatment. – Journ. Heredity 48, 47–50.

GAUL, H., 1958: Über die gegenseitige Unabhängigkeit der Chromosomen- und Punktmutationen. Zeitschr. f. Pflanzenzüchtg. 40, 151–188.

GELIN, O., L. EHRENBERG, and S. BLIXT, 1959: Quantitative studies of induced mutation in peas. II. Mutagenic effects of ethyleneimine and oxygen. – Agri Hort. Genet. 17, 265–274.

GIBSON, P. B., R. A. BRINK, and K. A. STAHMANN, 1950: The mutagenic action of mustard gas on *Zea mays*. – Journ. Heredity 41, 232–238.

GOLDACRE, R. J., A. LOVELESS, and W. C. J. ROSS, 1949: Mode of production of chromosomal abnormalities by the nitrogen mustards. – Nature 163, 667–669.

GUSTAFSSON, Å., 1947: Mutations in agricultural plants. – Hereditas 33, 1–100.

GUSTAFSSON, Å., and J. MAC KEY, 1948: The genetical effects of mustard gas substances and neutrons. – Hereditas 34, 371–386.

GUSTAFSSON, Å., and N. NYBOM, 1949: Colchicine, X-rays and the mutation process. – Hereditas 35, 280–284.

HAGBERG, A., 1960: Barley mutations used as a model for the application of cytogenetics and other sciences in plant breeding. – In press.

HAGBERG, A., Å. GUSTAFSSON, and L. EHRENBERG, 1958: Sparsely contra densely ionizing radiations and the origin of erectoid mutations in barley. – Hereditas 44, 523–530.

HESLOT, H. et R. FERRARY, 1958: Action génétique comparée des radiations et de quelques mutagènes sur l'orge. – Annales de l'institut national agronomique 44, 133–152.

HESLOT, H., R. FERRARY, R. LÉVY et C. MONARD, 1959: Recherches sur les substances mutagenes (halogéno-2-éthyl) amines, dérivés oxygénés du sulfure de bis-(chloro-2 éthyle), esters sulfoniques et sulfuriques. – Comptes rendus des séances de l'Académie des Sciences 248, 729–732.

KAPLAN, R., 1946 (1947): Über die Häufigkeit von Faktormutationen durch Röntgenbestrahlung des Pollens von *Antirrhinum majus* in verschiedenen Quellungszuständen. – Naturwiss. 33, 348–349.

KIHLMAN, B., and A. LEVAN, 1951: Localized chromosome breakage in *Vicia faba*. – Hereditas 37, 382–388.

KRONSTAD, W. E., R. A. NILAN, and C. F. KONZAK, 1959: Mutagenic effects of oxygen on barley seeds. – Science 129, 1618.

KREIZINGER, J. D., 1958: Diepoxybutane as a chemical mutagen in maize. – Maize Genetics Cooperation News Letter 32, 19–22.

LINNERT, G., 1953: Der Einfluß von Chemikalien auf Chiasmenbildung und Mutationsauslösung bei *Oenothera*. – Chromosoma 5, 428–453.

LOVELESS, A., 1951: Qualitative aspects of the chemistry and biology of radiomimetic (mutagenic) substances. – Nature 167, 338–342.

LOVELESS, A., and S. REVELL, 1949: New evidence on the mode of action of "mitotic poisons". – Nature 164, 938–944.

MAC KEY, J., 1954a: The biological action of mustards on dormant seeds of barley and wheat. – Acta Agr. Scand. 4, 419–429.

– 1954b: Neutron and X-ray experiments in wheat and a revision of the speltoid problem. – Hereditas 40, 65–180.

MARQUARDT, H., 1949: Quantitative Auswertung eines Versuches zur Auslösung von Chromosomenmutationen durch ein Äthylurethan/Kaliumchlorid-Gemisch. – Experientia V/11, 443.

MICHAELIS, P., 1958: Untersuchungen zur Mutation plasmatischer Erbträger, besonders der Plastiden. – Planta 51, 600–634, 722–756.

MOUTSCHEN-DAHMEN, J. et M. 1958a: Sur l'évolution des lésions causées par la 8-ethoxycafféine chez *Hordeum sativum* et chez *Vicia faba*. – Hereditas 44, 18–36.

– 1958b: L'action du myleran (di-méthane-sulfonyloxy-butane) sur les chromosomes chez *Hordeum sativum* et chez *Vicia faba*. – Hereditas 44, 415–446.

MOUTSCHEN-DAHMEN, M., J. MOUTSCHEN, and L. EHRENBERG, 1959: Chromosome disturbances and mutation produced in plant seeds by oxygen at high pressures. – Hereditas 45, 230–244.

MURPHY, C. F., and F. L. PATTERSON, 1958: The effects of the chemical mutagen, diepoxy-butane, on Clintland oats. – Proc. Iowa Acad. Sci. 65, 184–189.

NYBOM, N., 1954: Karyotype and viability in barley. – Acta Agr. Scand. 4, 507–514.

– 1956: On the differential action of mutagenic agents. – Hereditas 42, 211–217.

OEHLKERS, F., 1943: Die Auslösung von Chromosomenmutationen in der Meiosis durch Einwirkung von Chemikalien. – Zeitschr. f. indukt. Abstammungs- und Vererbungslehre 81, 313–341.

– 1949: Mutationsauslösung durch Chemikalien. – Sitzungsber. der Heidelberger Akad. der Wissenschaften, Mathematisch-naturwissenschaftliche Klasse, 9. Abhand. 1–40.

OEHLKERS, F., 1953: Chromosome breaks influenced by chemicals. – Suppl. to Heredity **6**, 95–105.
– 1956: Die Auslösung von Mutationen durch Chemikalien bei *Antirrhinum majus*. – Zeitschr. f. indukt. Abstammungs- und Vererbungslehre **87**, 584–589.

OEHLKERS, F., und H. MARQUARDT, 1950: Die Auslösung von Chromosomenveränderungen durch Injektion wirksamer Substanzen in die Knospen von *Paeonia tenuifolia*. – Zeitschr. f. indukt. Abstammungs- und Vererbungslehre **83**, 299–317.

RAPOPORT, I. A., 1948: Atsetilirovanie gennykh belkov i mutatsii. – Dokl. Akad Nauk SSSR **59**, 119–122.

RESPONDEK, V., 1958: Chemische Mutationslösung beim Mais. – Die Naturwissenschaften **45**, 555.

REVELL, S. H., 1953: Chromosome breakage by X-rays and radiomimetic substances in *Vicia*. – Suppl. to Heredity **6**, 107–124.

RÖBBELEN, G., 1959: 15 Jahre Mutationsauslösung durch Chemikalien. – Der Züchter **29**, 92–95.

SANDERS, M. E., C. J. FRANZKE, and J. G. ROSS, 1959: Influence of environmental factors on origin of colchicine-induced true-breeding diploid mutants in *Sorghum*. – Amer. Journ. Bot. **46**, 119–125.

SCHEIBE, A., and G. HÜLSMANN, 1958: Mutationsauslösung durch Chemikalien beim Steinklee (*Melilotus albus*). – Zeitschr. f. Pflanzenzüchtg. **39**, 299–324.

SSACHAROW, W. W., 1936: Jod als chemischer Faktor, der auf den Mutationsprozeß von *Drosophila melanogaster* wirkt. – Genetica **18**, 193–216.

STADLER, L. J., 1954: The gene. – Science **120**, 811–819.

STUBBE, H., 1932: Untersuchungen über experimentelle Auslösung von Mutationen bei *Antirrhinum majus*. III. — Zeitschr. f. indukt. Abstammungs- und Vererbungslehre **60**, 474–513.
– 1937: Der gegenwärtige Stand der experimentellen Erzeugung von Mutationen durch Einwirkung von Chemikalien. – Angew. Chemie **13**, 241–254.

WESTERGAARD, M., 1957: Chemical mutagenesis in relation to the concept of the gene. – Experientia XIII/6, 224–234.

WETTSTEIN, D. VON, Å GUSTAFSSON, und L. EHRENBERG, 1959: Mutationsforschung und Züchtung. – Arbeitsgemeinschaft f. Forsch. des Landes Nordrhein-Westfalen. **73**, 7–60.

WOLFF, S., 1957: Recent studies on chromosome breakage and rejoining. – Advances in Radiobiol., Edinburgh, 463–474.

Chemical mutagenesis as a tool in macromolecular genetics

M. Westergaard

Institute of Genetics, University of Copenhagen

The subject of this meeting is chemical mutagenesis and I have been asked to give a survey of the work on microorganisms.

Instead of trying to give a list of all the chemicals which have been used with positive or negative results in different species I should like to approach the subject in a somewhat different way. Rather than reporting what has already been done, I prefer to spend most of the time discussing what should and could be done in the future. Especially I want to discuss in some detail the methodological problems which must be solved in order to make an intelligent use of chemical mutagenesis in tackling some of the front line problems in modern macromolecular genetics. My excuse for introducing such a discussion in this paper is that only in microorganisms can we in my opinion expect to develop methods with sufficient resolving power to study genetic mechanisms on the macromolecular level.

Ever since 1927, when Muller established the mutagenic effect of X-rays by means of his famous ClB stock of *Drosophila,* the progress in experimental mutagenesis has largely depended upon the development of new and ingenious methods for scoring different types of mutations unambiguously on a quantitative scale.

In 1934 Timoféef-Ressovsky set up five rules to be followed in experimental mutation work. These are the following: 1. genetic purity of the material, 2. sufficiently large numbers in both controls and treated material, 3. suitable genetical methods for the detection of newly arisen mutations, 4. analysis of the detected variants as to their nature (whether non-heritable, cytoplasmic, chromosomal, genic), and 5. some knowledge of the manner in which the agent used can act on the germ cells of the treated organism (penetration; direct or indirect action).

These five rules are still as valid and indispensable in mutation work as they were in 1934, but today we can not only specify the rules more precisely, but we must also add additional specifications. Let me try to outline what I think are today the necessary requirements for such a system in the following 8 paragraphs:

The system should allow:

1. A quantitative estimate of "forward" and reverse mutations (spontaneous and induced) at specific loci.
2. Fine mapping of the forward mutations by means of a cross-over analysis of high resolving power.

3. A similar fine mapping of the reverse mutations to allow a distinction between back mutations and suppressor mutations. In the latter group ist must be possible to distinguish between suppressor mutations at different loci and intralocus suppressor mutations.
4. The recovery of the whole spectrum of mutations at a specific locus. A precise classification of the mutant alleles with respect to phenotypic expression is therefore necessary ("leaky" mutants, temperature-sensitive mutants, pH-sensitive mutants etc.).
5. An analysis of the gene controlled protein (determination of amino acid sequence).
6. The construction of an interallelic complementation map and a correlation of such a map with the fine crossover map.
7. An analysis of the sequence of events leading from the mutagenic treatment to the expressed mutation and a correlation of this sequence of events with the metabolic state of the treated cells (time of DNA replication, RNA and protein synthesis etc.). For all practical purposes this means working with a synchronously dividing system.
8. Finally it is important to have at least one such "ideal" system available in bacteria and/or viruses and at least one in higher organisms.

Some comments may be necessary to justify these ambitious requirements.

1. The necessity for studying "forward"[1]) *and reverse mutation rates at specific loci*

Classical genetic theory is based upon experiments in which it was possible to handle 10^3–10^5 individuals in crossing-over and mutation analysis. This resolving power was high enough to study crossing-over between genes, but not interallelic recombination, and in mutation experiments, the resolving power was high enough to obtain data on mutation rates of a broad spectrum of mutations (*e. g.* lethals along the X-chromosome of *Drosophila*), but, with few exceptions, it was insufficient to allow a quantitative estimate of mutation rates at specific loci.

The transition from classical genetics to macromolecular genetics was made possible because the resolving power of the recombination analysis was increased to $1/10^7$(cf. Pontecorvo, 1958) which brings it to the macromolecular level. In a few microorganisms it is now possible to do a mutation analysis with a similar resolving power.

In recombination analysis as well as in the analysis of mutations this increase in resolving power has been possible only by using some kind of screening methods which makes it possible to select for very rare specific genetic events. In mutation work especially, this imposes a very serious problem namely to select for mutations at specific loci without at the same time introducing interallelic selection. This problem has not been solved in the various ingenious "filtration" or "enrichment" methods which are widely used in mutation work in bacteria (cf. Lederberg, 1950) and fungi (cf. Woodward, *et al.*, 1954). By such methods, interallelic selection is probably unavoidable (for instance, most "leaky" mutants will be lost), and in my opinion these methods should not be used in quantitative mutation experiments. (*cf.* § 4).

[1]) The term "forward" mutation is a hideous one. It is used as a counterpart to "back" or „reverse" mutation. In most cases a „forward" mutation leads to a loss of function and a reverse mutation means a restoration of function. Restoration of function may be caused by a backmutation or by a suppressor mutation.

2. The necessity to determine the position of the individual mutant alleles within the controlling region of the chromosome [1])

"Fine mapping" requires a recombination analysis with very high resolving power, and preferably also with closely linked outside markers on both sides of the gene. It is therefore necessary to be able to select for very rare crossover events. PRITCHARD, 1957, and BENZER, 1957, should be consulted for such systems (see also PONTECORVO, 1958).

3. It is necessary to characterize the individual mutant alleles with respect to their stability

Some mutants are stable, others may undergo reverse mutations; spontaneously and/or after mutagenic treatment. If they do undergo reverse mutations, we must then be able to decide in appropriate crosses if the reverse mutation is due to a back mutation or to a suppressor mutation. In the latter case, that of reversion by suppressor mutation, we must be able to map the position of the suppressor, and since cases are known of intragenic suppressor mutations (cf. STARLINGER *et al.*, 1956), this again requires a very high resolving power in the recombination analysis. BENZER, 1957, FREESE, 1959 and STARLINGER *et al.*, 1956 should be consulted for the few cases in which such an analysis has been possible.

Finally we must be able to characterize the suppressor mutation with respect to its specificity. It may be allele specific, locus specific, or of a more broad spectrum type. If the gene controlled protein is an enzyme, we can expect that it will soon be necessary to correlate the action of the suppressor with the function of the enzyme controlling system (cf. § 5).

4. A precise characterization of the mutant alleles with respect to phenotypic expression

There is a great variation with respect to phenotypic expression between mutants at a single locus. In some mutants the normal function of the gene is only slightly changed or impaired and at the other end of the spectrum we find mutants with complete loss of function. The first group of mutants may often be of great interest, but they are the ones which escape detection most easily.

5. Analysis of gene function

Our concept of the primary function of the gene has undergone a rapid evolution during the last years. The "one gene – one enzyme" hypothesis of BEADLE and TATUM has been specified in the "sequence hypothesis" (cf. CRICK, 1958), which states that a certain base sequence in the DNA (or RNA) molecules determines, that is, codes for, a certain amino acid

[1] (the "cistron" in BENZER's terminology, *cf.* BENZER, 1957).

sequence in the gene controlled protein, the hypothesis implying that the specificity of the protein is determined by its amino acid sequence.

The experimental evidence for this "sequence hypothesis" stems exlusively from the beautiful work of INGRAM, ITANO, PAULING and others (cf. HUNT and INGRAM, 1958) on the structure of the human hemoglobin proteins. (It is worth noticing that one of the most important new developments in genetics comes from human genetics!)

According to the sequence hypothesis at least some intralocus point mutations are the results of changes in the base sequence in DNA (or RNA), the new sequence coding for a different amino acid sequence in the protein.

To establish a correlation between a fine mutation map, constructed by means of mutations induced and reverted by mutagens of known chemical affinity to the purine and pyrimidine bases on one hand, and the amino acid sequence of the gene controlled protein on the other hand, seems today one of the few promising experimental approaches to one of the fundamental problems of macromolecular genetics, that of "breaking the code". This problem has hitherto been tackled only as an abstract mathematical problem (cf. GOLOMB *et al.*, 1958 for further references; see also DOTY *et al.*, 1959).

If the gene controlled protein is an enzyme, it may soon not be enough to have isolated the enzyme, but we will also need to have the whole enzyme controlling system worked out, biochemically as well as genetically. The work of MONOD and coworkers on β-galactosidase should be consulted for such a thoroughly analysed enzyme system (cf. PARDEE, 1959). We should not forget the hypothesis so often expressed by GOLDSCHMIDT (cf. GOLDSCHMIDT, 1955) that there are two main types of gene function; some genes control enzyme specificity and others control the rate of enzyme formation (cf. also the work of HOROWITZ and coworkers on the tyrosinase system of *Neurospora* (HOROWITZ, *et al.*, 1959)).

6. Complementation

During the last 2–3 years a new genetic phenomenon has been discovered, namely that of interallelic complementation. Sometimes mutants derived from the same functional region of the chromosome have been found to complement each other; when combined, for instance in a heterocaryon, a partially wild type phenotype is produced. The phenomenon of interallelic complementation seems to be restricted to some loci, and is absent in others. When interallelic complementation is found it has, in a few cases, been possible to construct a complementation map, and to correlate this map with the fine crossing-over map. The degree of complementation, *i. e.* the fraction of wild type activity found in the heterocaryon, seems to increase with the distance between the alleles within the locus, but the activity never exceeds 25 per cent of the wild type (cf. GILES, 1959). The recent discovery by WOODWARD (1959) in *Neurospora* of interallelic complementation *in vitro* suggest that we are dealing with a mechanism which operates on the protein level.

The new phenomenon of interallelic complementation gives us one more operational possibility for characterizing individual mutant alleles at the same locus. The actual mechanism which gives rise to the formation of what has been called "hybrid protein" (cf. SINGER and ITANO, 1959) is still rather obscure. Our ideal mutation system should also be able to cope with an analysis of this phenomenon which undoubtedly will be a central problem in genetics for a long time. Again we must be able to isolate the gene controlled protein in a highly purified state in order to solve the problem.

7. *Correlating the different steps involved in the induction of a mutation with the metabolic state of the treated cells*

The analysis of chromosome breakage in plants and animals has revealed that several steps are involved in the production of an actual break. A distinction is made between "potential" breaks and "actual" breaks, and in a number of cases it has been possible to discriminate experimentally between the two categories. It is known that the great majority of the "potential" chromosome breaks undergo restitution, and that a small fraction only are seen as actual breaks. Whether a potential break becomes an actual break or undergoes restitution depends upon the metabolic state of the cell during a short period after the mutagenic treatment (and probably upon the type of mutagen used to induce chromosome breakage).

This distinction between potential and actual mutations is important in all mutation work. However, with the exception of the pioneer work of WITKIN, the problem has so far been somewhat neglected in microorganisms (cf. WITKIN, 1959, see also DOUDNEY and HAAS, 1959). In bacteria, WITKIN has described two groups of UV induced mutations. One group, comprising mutations from auxotrophy to prototrophy is expressed only when protein synthesis takes place in the treated cells during the first hour after irradiation. If protein synthesis is inhibited, for instance with chloramphenicol, the mutations are not expressed. Another group of mutations, from streptomycin sensitivity to streptomycin resistance, is expressed irrespective of whether protein synthesis takes place after irradiation or not. My only criticism of these important experiments is that they do not quite live up to the standard of TIMOFÉEF-RESSOVSKY's fourth rule (cf. p. 30) because the mutants have not been analysed by an appropriate genetical crossing analysis.

Despite this reservation I think WITKIN has shown that the primary effect of a mutagen is to induce some kind of metastable state (a "promutation" in her own terminology), and the metabolic state of the cells after the treatment decides whether this "potential" mutation shall become an actual mutation or undergo restitution. The fact that different types of mutations show different patterns indicates that more than one mechanism must be involved.

However, the sequence of events leading to an actual mutation depends not only upon the type of mutation which is induced but also upon the mutagen used, because, as shown recently by BRENNER and SMITH (1959) and by LITMAN and PARDEE (1959) with phage T2, the mutagenic effect of 5-bromouracil is not inhibited when protein synthesis in the host bacteria is blocked by chloramphenicol.

In order to understand the mechanism (or rather mechanisms) of mutation, much more information is needed correlating the induction and expression of "forward" as well as of reverse mutations, induced by different mutagens, with the metabolic state of the cells, during and after treatment. It is most important to establish a possible correlation between the transition from "potential" to "actual" or "restituted" states with the time of DNA replication, of RNA synthesis, of protein synthesis and of energy turnover in the cells. There are many technical difficulties involved in such experiments, and this line of investigations can undoubtedly be pursued best by working with cell populations which divide synchronously. Following the pioneer work of ZEUTHEN and coworkers on *Tetrahymena*, various methods are now available for growing bacteria in synchronously dividing cultures (cf. MAALØE, 1959). Similar possibilities should exist in yeast and in unicellular algae. However, such systems have not yet been used systematically as tools in experimental mutagenesis.

8. The necessity for working with species on different evolutionary levels of organization

Although bacterial and viral genetics are hardly more than 20 years old, we are today in the rather awkward position that on the macromolecular level we seem to know much more about the genetic mechanisms in bacteria and viruses than in higher organism.

This is because we have a good deal of information about the structure and replication of the genetically important macromolecules in bacteria and viruses, whereas we know far less about the macromolecular structure of the genetically important structure in higher organisms, the nucleo-protein chromosome.

The transformation experiments have proved that the genetical information in bacteria is in the DNA (AVERY *et al.*, 1944). In bacterial viruses, the genetic information is in DNA (HERSHEY and CHASE, 1952); in TMV it is in RNA (*cf.* FRAENKEL-CONRAT et al., 1957). We know a good deal both about the macromolecular structure and the mechanism of replication of DNA (*cf.* Crick, 1957, DELBRÜCK and STENT, 1957), and we are also beginning to know something about different types of RNA (*cf.* WATSON, 1957, CRICK, 1958).

In higher organisms the bulk of genetic information is undoubtedly carried in the chromosomes; we know that the chromosome is made up of nucleo-proteins and that it contains both DNA and RNA. Although a number of ingenious models have been proposed (of which the one of FREESE's has some appeal to the present author; cf. FREESE, 1958), we have to admit that very little is actually known about the macromolecular structure of the chromosome or of its mode of replication. Am I too pessimistic when I say that it is justified to speak about macromolecular genetics only in bacteria and viruses (and in man!)?

The remarkable success with which microbial genetics has been able to explain some of the basic mechanisms like replication and recombination and to some extend even the mechanism of mutation on a macromolecular level has had a tremendous impact upon our thinking about similar mechanisms in higher organisms, and I believe that it has even divided geneticists into two schools, which we may call the "Generalizationists" and the "Evolutionists".

The first school generalizes from bacteria and viruses to higher organisms and believe that the fundamental genetical mechanisms: replication, recombination and mutation, are the same in bacteria, viruses and man (note that I put viruses after bacteria; this is because the virus, being a parasite, must be placed after bacteria on the tree of evolution!). PONTECORVO (1958) has thus recently rejected the classical chiasmatypy theory of crossing-over being due to reciprocal exchanges between segments of homologous chromosomes at meiotic pachytene. He proposes the "copy-choice" mechanism (*cf.* LEDERBERG, 1955), which may be the only mechanism of recombination in bacteria and viruses, as the universal mechanism of intra-chromosomal recombination.

I think that this school is sometimes inclined to neglect the most important of all biological theories, the theory of evolution, and to forget that the most important steps in evolution may be those which lead to new genetic mechanisms. DARLINGTON was the first to attempt an analysis of these steps in his masterly monograph of 1939 (cf. also DARLINGTON, 1959).

The "Evolutionist" school, which may be in the minority, is perhaps more impressed by some of the differences between the genetic mechanisms of bacteria and higher organisms than by the similarities, and they believe that these differences may reflect some important steps in the evolution of genetic mechanisms which mark the transition from the bacterial cell to the cell of higher organisms. Let me list some of the differences which have impressed me:

1. In bacteria (and viruses) the transfer of information from one cell to another seems always to be fractional. In sexual recombination, fertilization is "merozygotic" (WOLLMAN *et al.*,

1956, *cf.* also Jacob 1958), *i. e.* only a piece of the parental DNA (from the *Hfr* gamete) becomes incorporated into the zygote. When transfer of information is effectuated by transduction (cf. Hartman, 1957) or transformation, even a much smaller fraction of the genome of the donor cell is incorporated into the receptor cell.

In higher organisms, fertilization is "holozygotic", *i. e.* the whole male genome is incorporated into the zygote. Partial transfer of the genome of one cell into another seems to be a very rare phenomenon in higher organisms, the old "burdo" experiments of Winkler (1935) may perhaps be cited as an example. Furthermore, although a number of attempts have been made with various organisms, transformation or transduction experiments with higher organisms have not given convincingly positive results.

2. In bacteria and viruses, genetic recombination between homologous pieces of DNA seems to be effectuated by a non reciprocal "copy-choice" mechanism (Lederberg, 1955). By definition, such a mechanism must operate at the time of DNA replication. In higher organisms, recombination between homologous chromosomes – crossing-over – is normally reciprocal, it takes place during meiosis at the time when the homologous chromosomes are regularly paired (at pachytene).

Experimental evidence, obtained by several independent methods, leads to the conclusion that chromosome duplication (DNA, as well as protein replication) takes place during interphase prior to leptotene (cf. Taylor, 1957, 1959). I therefore agree with Taylor that we are forced to reject the classical Belling-Darlington hypothesis that crossing-over takes place at the time of chromosome duplication. Admittedly, this unfortunately means that at the moment we must admit absolute defeat with respect to understanding the actual mechanism of crossing-over. However, the fact that we fail to understand the mechanism of a biological phenomenon does not necessarily mean that we have to deny the reality of the phenomenon.

I think that this much discussed conflict can now be stated quite clearly. If one believes that the same recombination mechanism is operating in all species from bacteria to man, it is necessary to reject either – as Darlington does – the experimental evidence for interphase (pre-leptotene) DNA replication in higher organisms, or to reject – as Pontecorvo does – all the experiments and observation on which the classical theory of reciprocal pachytene crossing-over is based. The alternative is to accept two different mechanisms, copy-choice as the only one operating in bacteria and viruses, and reciprocal pachytene crossing-over operating as the normal mechanism of recombination in higher organisms. Personally I believe in two mechanisms, the second one representing an important step in the evolution of holozygotic sexual organisms.

We also know that another recombination mechanism is operating as a rare event in higher organisms. It has been called "gene conversion" and many other names, but I believe that it is the copy-choice mechanism of bacteria which is carried over as a kind of relic mechanism in higher organisms, in which it is normally replaced by reciprocal pachytene crossing-over (*cf.* Freese, 1957). I would not be surprised if it turns out that copy-choice recombination in higher organisms is mainly restricted to intralocus recombination. Since copy-choice recombination also in higher organisms must take place at the time of DNA replication, copy-choice recombination and reciprocal crossing-over in higher organisms must be separated in time. However, I do not think that some of the critical experiments which might prove or disprove this hypothesis have yet been done (*cf.* my discussion of Roman's paper in Roman and Jacob, 1958). It still remains to be seen, if one, or the other, or both mechanisms are operating in somatic crossing-over in higher organisms (*cf.* Pontecorvo, 1958, Roman and Jacob, *l. c.*).

3. In higher organisms the genetic information must be organized in an one dimensional structure. This can be microscopically identified as the nucleo-protein chromosome. The chromosomes undergo a very precise division cycle during cell division (mitosis), and a very

complex mechanism, meiosis, exists during which chromosome segregation and crossing-over take place.

In bacteria we know that the genetical information is likewise organized in a one dimensional structure which is also endowed with a very precise mechanism of replication. However, this structure has not been identified microscopically and I have not yet seen any convincing demonstration of mitosis or meiosis in bacteria, although claims for such observations have been made (*cf.* Spooner and Stocker, 1956). The present uncertain status of bacterial cytology is due to the smallness of the cells. I suggest that mitosis and meiosis have not been seen in bacteria simply because they do not possess such mechanisms.

The differences between bacteria and higher organisms outlined in these three paragraphs may all fit a hypothesis, which consideres a "naked" nucleic acid chromosome as the information carrying structure in bacteria and viruses, and the nucleo-protein chromosome as the information carrying structure in higher organisms.

The difference between the "naked" nucleic acid chromosome and the nucleo-protein chromosome would be that of a much more intimate connection between nucleic acid and protein in the latter than in the former.

This hypothesis does not necessarily require that any genetic information is carried in the protein of the nucleo-protein chromosome, because the function of the protein may mainly be to stabilize and protect the information carrying molecules.

This hypothesis leads to the following conclusions about differences between bacteria and higher organisms with respect to fundamental genetical mechanisms:

In bacteria and in viruses the stability and preciseness with which genetic information is transmitted from cell to cell is effectuated by the stable and precise mechanism of complementary replication of the DNA molecule (*cf.* Delbrück and Stent, 1957, Meselson and Stahl, 1958).

In bacteria, when fragments of DNA from a donor cell are introduced into a receptor cell, either by cell fusion – as in the *Hfr*/*F*⁻ system – or by transduction and transformation, incorporation and genetic recombination is effectuated by a copy-choice mechanism at the time of DNA replication. Recombination in phage, and recombination between phage chromosomes and bacterial chromosomes (in processes like lysogenization, cf. Bertani, 1958, Luria *et al.*, 1958), would be effectuated through a similar copy-choice mechanism.

In higher organisms, it is the precise mechanism of mitosis which controls the division of the nucleo-protein chromosomes. Mitosis requires the evolution of spindle apparatus, of centrioles and of kinetochores, structures which all seem to be absent in the bacterial cells. In higher organisms, the highly sophisticated and refined mechanism of meiosis takes care of genetic recombination through chromosome segregation and reciprocal crossing-over, copy-choice recombination, however, still being preserved as a rare relic phenomenon, but transfer of information by DNA infection (transduction, transformation) no longer possible.

We may perhaps add one more feature to the list of differences between the bacterial cell and the cells of higher organisms. I think that the latter are characterized by a much more marked division of labor between nucleus and cytoplasm than the former. The fact that bacterial nuclei apparently lack a nuclear membrane lends support to this view (*cf.* Kellenberger *et al.*, 1959, Birch-Andersen and Maaløe, 1955).

This implies that the rather clearcut distinction between nuclear inheritance and cytoplasmic inheritance which can in most cases be drawn in higher organisms may not be nearly as clearcut in bacteria.

My reason for proposing this very oversimplified model here is not only that I want to emphasize the importance of keeping the evolutionary aspect in mind when comparing the genetic mechanisms of bacteria with those of higher organisms, but also because I think that the use of chemical mutagens is one possible way of testing the validity of the hypothesis experimentally. I have, perhaps prematurely, drawn the conclusion from the mutation work on bacteria that the bacterial cells seem to be much more susceptible than the cells of higher organisms to mutagenic treatment, and especially to mutagens which are likely to react directly with the purine and pyrimidine bases, thereby messing up base pairing and DNA replication. Although I admit that far too little work has been done with such mutagens in higher organisms, the preliminary data may fit the hypothesis that the purine and pyrimidine bases in the nucleo-protein chromosomes are better protected from outside agents than they are in the "naked" nucleic acid chromosomes.

I have spent most of the time alloted to me in this lecture writing out the specifications of the ideal mutation system. This should enable us to study possible correlations between the fine crossover map, the complementation map, the stability, expressivity and function of mutant alleles, with the mutagens used to induce mutations and with the conditions under which mutagens are applied. I have in addition spend – perhaps too long a time – putting in a plea for such work in organisms on different levels of organization. I have seen so many geneticists switch from higher organisms first to fungi or algae than to bacteria or phage, but very few have climbed the tree of evolution in the opposite direction. Personally I do firmly believe that not all problems in genetics can be solved by studying phage.

I should like to round up the theoretical part of this paper with a few remarks about the selection of mutagens to be used in such ideal mutation systems.

One of the first chemical mutagens, mustard gas, was tested on *Drosophila* because of its general radiomimetic effect, and this discovery led to the search for chemicals with similar properties and the family of chemicals which are known as the alkylating agents came to be widely used in chemical mutagenesis, especially in higher organisms, as you have heard in the two preceding lectures. Other chemicals were tested for mutagenic effect because they were believed to play an important role in radiation chemistry. Hydrogen peroxide and organic peroxides belong to this group. Still other compounds were tested because they had possibilities as tumour inhibitors in cancer chemotherapy. To this group belong not only many alkylating chemicals, but also some antimetabolites and enzyme inhibitors. Quite a few mutagenic chemicals have been discovered simply by trial and error. Finally a group of chemicals have been tested because they were predicted to either interfere with base pairing or have direct chemical effect on the purine and pyrimidine bases. This group includes base analogues like 5-bromo-uracil (LITMAN and PARDEE, 1956), and nitrous oxide. This group of compounds has mainly been tested in bacterial and viral systems.

It must be admitted that the alkylating agents which are the mutagens most widely used in mutation work in higher organisms may be as unspecific, and therefore as unsatisfactory analytical tools in macromolecular genetics, as high energy radiation has proved to be. At the moment, most of the interesting results in chemical mutagenesis come from the work with mutagens which we hope can be called base specific. It therefore seems very important to study the mutagenic effect of these chemicals much more systematically in suitable mutation systems in higher organisms than has hitherto been done. On the other hand, I would also put in a plea for more work with the alkylating agents in bacteria and viruses, because a good deal of information about these agents has already been accumulated in higher organisms, and in our own mutation system, some of the alkylating agents have a remarkably specific mutagenic effect (cf. table 1 und 2).

Table 1. The mutagenic effect of a series of homologous alkyl-methanesulphonates on the reverse mutation rate of the double mutant adenineless, inositolless of *Neurospora crassa* (G. Kølmark, unp.; preliminary data)

Mutagen	Time treated with 0,1 Mol.	per cent survived	Mutations per 10^6 ad$^+$	treated conidia inos$^+$
$BrCH_2CH_2OSO_2CH_3$	2 hours	52	152	.04
$ClCH_2CH_2OSO_2CH_3$	10 —	56	88	.3
$FCH_2CH_2OSO_2CH_3$	20 —	90	10	.16
$CH_3CH_2OSO_2CH_3$	12.5 —	14	17	11
$CH_3OSO_2CH_3$	7 minutes	90	60	.1

Table 2. The relative specificity of mutagens on the backmutation rate of two alleles, adenineless, inositolless, combined in a double mutant of *Neurospora crassa*. The mutagens are compared under "optimal conditions", which are defined as the conditions of treatment (concentration of mutagen, duration of treatment etc.) which give the highest overall number of mutations per 10^6 treated macroconidia. Note that we are dealing with an allele specific effect, and not with a locus specific effect. (Data compiled from published and unpublished experiments of G. Kølmark, cf. Westergaard, 1957).

Mutagen	Mutations per 10^6 macroconidia inos$^+$	ad$^+$	Proportion ad$^+$/inos$^+$
1. Spontaneous rates	.02	.2	10
2. Ethyl methane sulphonate	11.3	17.4	1.5
3. Diethylsulphate	4.3	16.8	4
4. Dimethylsulphate	3.4	84.0	19
5. Chloroethyl methane sulphonate	.3	51.0	190
6. Diepoxybutane	.2	89.0	445
7. Bromoethyl methane sulphonate	.04	152.0	3800
8. Ultraviolet light	7.1	3.5	.5
9. X-rays	.2	3.2	16

Some of you may think that I have been daydreaming about the ideal mutation system, and that such a system will not be found. Let us now see how close the mutation work on microorganisms has brought us to the ideal system, keeping in mind that we want at least one such system in bacteria and/or viruses, and at least one in higher organisms.

To me there is no doubt that at present the rII mutation system in phages of *E. coli* which has been worked out so beautifully by Seymour Benzer comes closest to our ideal (*cf.* Benzer, 1957). In a separate contribution during this symposium Dr. C. Bresch will explain how this system works. It is possible in Benzer's system to score forward and reverse mutations with very high accuracy (the resolving power is appr. $1/10^7$ or higher). It is possible to do fine mapping of the mutants with a similarly high accuracy. It is possible to classify the mutants with respect to phenotypic expression and it is possible to study complementation. Synchronously dividing cultures of the host cells can be obtained, although this approach has, as far as I know, hitherto been somewhat neglected.

Actually Benzer's system matches 7 of the 8 requirements which I have specified, and it has only one shortcoming: it has not been possible to isolate and purify the gene controlled protein. So much information has now been obtained with respect to mutational sites ("hot spots") etc. in this system that the gene controlled protein is really badly needed. As far as I understand it, the possibility of isolating the "rII protein" is evaluated rather pessimistically, and therefore, despite all of its beautiful refinements, the rII mutation system may not be the ideal system.

Realizing this, a number of geneticists are now looking for other systems, which not only have the genetical refinements of BENZER's system, but in which the protein can also be analysed. Their approach to the problem is most interesting. Realizing that the stumbling block in the attempts to make BENZER's system perfect was the failure to isolate the protein, they have decided to start from the "protein end". They now look, in viruses and bacteria, for good proteins which can be isolated and purified sufficiently to establish the number of polypeptide chains and the amino acid sequence. Not until they have found such a protein do they begin to worry about the genetical aspects of the problem (quantitative estimates of forward and reverse mutations, fine mapping etc.). At the moment the most promising attempts along such lines seem to be the work of LEVINTHAL and GAREN on the alkaline phosphatase of *E. coli*, the work of BRENNER, STREISINGER and others on the "head" and the "tail" proteins of phage (Dr. LEVINTHAL, personal communication).

In bacteria and viruses, possibilities for finding the ideal mutation system in the near future thus seem rather good. In higher organisms the situation is less developed and it will probably take many years before we can catch up with the phage geneticists. Both in *Neurospora*, *Aspergillus* and in yeast some fine mapping of interallelic mutants has been done. In some cases it is also possible to construct a complementation map, and in a number of cases it seems possible to purify the gene controlled protein (cf. FINCHAM, 1959). However, in none of the loci, in which a fine mapping is possible or in which an enzyme is known, have methods been worked out for scoring both forward and reverse mutation rates quantitatively. In the very few cases in fungi in which it is possible to obtain quantitative data on mutations in both directions, it has not yet been possible to do a fine mapping, and/or the protein is not known. I believe that at the moment the *pan*-2 locus of *Neurospora* (cf. MARY CASE and GILES, 1958) is the most promising one because it offers the best opportunities with respect to fine mapping, complementation mapping, and isolation of the protein (Dr. GILES, personal communication). But – alas – a screening system for forward mutations has not been designed.

We have in our own laboratory for many years worked with a quantitative, locus specific, back mutation system in *Neurospora*, realizing the importance of obtaining data on forward mutation rates at the same specific locus. We are therefore working on a quantitative, locus specific forward mutation system which has been recently developed by DESERRES and KØLMARK at Oak Ridge. They have generously placed this system at our disposal, and we are now just beginning to collect data, using, at present, alkylating chemicals as mutagenic agents. The method of DESERRES and KØLMARK takes advantage of the fact that all adenine requiring *Neurospora* mutants which belong to the *adenine*-3 locus have a bright purple colour. They have constructed a properly marked, microconidial, colonial strain which can be grown in big flasks containing, in our experiments, appr. 8 liters of substrate, with appr. one million colonies in each flask. After a mutagenic treatment, conidia are grown in such flasks for appr. 5 days, then the colonies are poured out in a big white pan in which the purple colonies can be detected and picked up easily, and without ambiguity. The resolving power of the system is high enough to establish a quantitative relationship between mutagenic treatment and mutation yield in X-ray experiments as shown by the fact that DE SERRES and KØLMARK (1958) have found a "one hit" curve after X-ray treatment. The *adenine-3* locus of

Neurospora is a compound locus covering two adjacent cistrons, *adenine-3* A and *adenine-3* B. By a conventional heterokaryon test it is possible to refer the individual purple mutants to the A- or B-cistron, and the X-ray experiments quoted above have shown that the A-cistron is less sensitive to X-rays than the B-cistron. The reverse mutation rates of these adenine mutants can be scored easily by means of our usual back mutation technique (*cf.* WESTERGAARD, 1957).

In higher organisms the *adenine-3* system is probably the best for studying forward and reverse mutations at a specific locus quantitatively. Unfortunately the system has several drawbacks. The high sterility of interallelic crosses between adenine mutants makes fine mapping difficult and – alas – again the gene controlled protein, the enzyme, is not available.

Another quantitative "forward-reverse" mutation system has been constructed in *Neurospora* (REISSIG, 1959), but only a preliminary note has been published, and as far as I know, the system has not yet been used in chemical mutagenesis.

Even the best mutation systems in filamentous fungi may end up with one shortcoming, and the more refined the systems become in other respect, the more serious this may be. Sooner or later the problem will come up of relating the induction of mutations with the metabolic state of the cells, a problem which I believe is best studied by using synchronously dividing cultures. I am rather pessimistic about the possibility of obtaining synchronously dividing cultures in filamentous fungi, and let me therefore close this part of the paper by predicting that the ideal mutation system in higher organisms will not be developed in filamentous fungi. It may be found in yeast, but more likely in unicellular algae like *Chlamydomonas*. Therefore, the sooner we begin to look for good proteins, controlled by a mutation system of sufficient sophistication, in such organisms, the better are our chances for catching up with the bacterial and viral geneticists.

I consider it my duty also to speak briefly about some of the results which have been obtained with some of the mutation systems now available in microorganisms, but for obvious reasons I shall confine myself to the results obtained with the few systems which come closest to my ideal. The recent beautiful work of the Tübinger group on mutations in TMV are omitted here only because I know that Professor MELCHERS will tell about it in a separate communication.

In BENZER's rII system, BENZER and FREESE (1958), BRENNER et al. (1958), and FREESE (1959*a*, *b*) have done a fine mapping of spontaneous mutations, and of mutations induced by two base analogues, viz. 5-bromo-uracil or 5-bromo-deoxyuridine (BD) and 2-aminopurine (AP), and by proflavine. The results have recently been summarized (FREESE, 1959*a*, *b*). His graph marks in my opinion the present culmination in the successful attempts to use chemical mutagens as an analytical tool in macromolecular genetics.

The results can be briefly summarized as follows: The four mutagens give different "hot spots" on the fine map, and the position and height of these "hot spots" differ to some extent from the spontaneous ones. Mutations induced by AP and BD can also be induced to revert by these compounds, but AP and BD cannot induce reverse mutations in spontaneous or proflavin induced mutations, although they will undergo spontaneous backmutations.

FREESE uses these data to discuss two possible mechanisms of mutation in terms of base pairing. In one mechanism, a purine (or pyrimidine) base is replaced by another purine (or pyrimidine) base, in the second mechanism a purine base is replaced by a pyrimidine base, or *vice versa*. Without going into details about this hypothesis I want to emphasize that it is now possible to discuss the mutation mechanism intelligently on a macromolecular level, using chemical mutagens as the analytical tool.

With respect to the work of higher organisms (fungi and algae) I have little to say. Our own work has hitherto been confined to the study of chemically induced back mutations in a double mutant of *Neurospora*, and we have mostly used alkylating agents as mutagens (cf. WESTERGAARD, 1957). We are just now beginning to collect data on forward mutations by means of the DESERRES-KØLMARK *adenine-3* system, and Dr. MALLING in our laboratory has already recovered appr. 300 purple adenine mutants induced by alkylating chemicals; these mutants are now being analysed.

Concluding remarks: In my opinion, chemical mutagenesis is today facing two main problems. One is to develop "the ideal mutation system" in species on different evolutionary levels of complexity. I believe that geneticists and protein chemists together will soon have solved this problem in viruses and in bacteria, and I think that we will also in due time have such systems in higher organisms.

The second problem is to make an intelligent use of mutagenic chemicals as analytical tools in macromolecular genetics. I am dutifully impressed by the beautiful work with the base analogues in bacteria and virus, but I would not be surprised if it turns out that they do not always act as specifically as we believe today. The living cell is such an intricately balanced system and such an unpleasantly complex substrate to treat with chemicals, and it is unfortunately only in a RNA virus (TMV) – not yet in the DNA virus, that we can take out the nucleic acid, treat it in vitro, put it back again, and look for mutations! [1])

So far, most of the attempts to use "nucleic acid specific" mutagens have been confined to chemicals which may interfere with base pairing. There are good and obvious reasons for this, but perhaps too little attention has nevertheless been paid to the "backbone" of the nucleic acids (sugar and phosphoric acid). I am undoubtedly influenced by our own work on specific backmutations induced by various alkylating agents. We find a remarkable degree of relative specificity in the backmutation pattern of two alleles, although we use mutagens which one would expect had a rather unspecific affinity to many constituents of the cells. I can not help thinking that there may be some leads in these experiments, and that we should in this case think in terms of "backbone chemistry" and how changes in the backbone – and in the protein – may effect the stability of the base sequence. However, unfortunately we are just not good enough chemists to see the light.

May I therefore close this lecture with one more plea: Almost from the beginning physical mutagenesis profited greatly by the fact that many physicists (DELBRÜCK, ZIMMER, LEA, FANO and others) immediately realized that here was an interesting and important new field which needed the cooperation between geneticists and physicists with a biological outlook. Surprisingly enough, chemists and especially physical chemists have until now hardly discovered that in chemical mutagenesis there is a borderline field which provides fascinating biological and chemical problems. I believe that the main problem which chemical mutagenesis is facing today is one of communication: to establish as successful a collaboration between geneticists and chemists as was established a long time ago between geneticists and physicists.

Aknowledgements

This work was supported by grants from the Carlsberg Foundation and from The Danish Science Foundation. The author is indebted to Dr. Ruth SAGER for correcting the manuscript and for valuable suggestions.

---

[1]) Note added in proof: This has now been done also with the TP, *cf.* R. M. LITMAN and H. EPHRUSSI-TAYLOR, 1959: Compt. Rend. Acad. Sci. (Paris), **249**, 838–840.

Literature

[1] AVERY, O. T., MACLEOD, G. M., and MCCARTY, M., 1944: Studies on the chemical nature of the substance inducing transformation of pneumococcal types. Induction of transformation by a desoxyribonucleic acid fraction isolated from *Pneumococcus*, type III. – J. Exp. Med. **79**, 137–158.

[2] BENZER, S., 1957: The elementary units of heredity. In "The Chemical Basis of Heredity", pp. 70–93. Johns Hopkins Press, Baltimore.

[3] BENZER, S., and FREESE, E., 1958: Induction of specific mutations with 5-bromouracil. – P. N. A. S. (Wash.) **44**, 112–119.

[4] BERTANI, G., 1958: Lysogeny. – Adv. Virus Res. **5**, 151–193.

[5] BIRCH-ANDERSEN, A., and MAALØE, O., 1953: High resolution electron micrographs of sections of *E. coli*. – Biochimica et Biophysica Acta 12, 395–400.

[6] BRENNER, S., BENZER, S., and BARNETT, L., 1958: Distribution of proflavin induced mutations in the genetic fine structure. – Nature 182, 983–985.

[7] BRENNER, S., and SMITH, J. D., 1959: Induction of mutations in the desoxyribonucleic acid of phage T2 synthetized in the presence of chloramphenicol. – Virology 8, 124–125.

[8] CASE, MARY, and GILES, N. H., 1958: Recombination mechanisms in the *pan-2* locus of *Neurospora crassa*. – Cold Spring Harbor Symp. Quant. Biol. **23**, 119–136.

[9] CRICK, F. H. C., 1957: The structure of DNA. – In "The Chemical Basis of Heredity", pp. 532–549. Johns Hopkins Press, Baltimore.

[10] CRICK, F. H. C., 1958: On protein synthesis. – Symp. Soc. Exp. Biol. 12, 138–163.

[11] DARLINGTON, C. D., 1939: The Evolution of Genetic Systems. – Cambridge Univ. Press. (2nd edition, 1959).

[12] DELBRÜCK, M., and STENT, G., 1957: On the mechanism of DNA replication. – In "The Chemical Basis of Heredity", pp. 699–735. Johns Hopkins Press, Baltimore.

[13] DE SERRES, F. J., and KØLMARK, H. G., 1958: A direct method for determination of forward-mutation rates in *Neurospora crassa*. – Nature 182, 1249–1250.

[14] DOUDNEY, C. O., and HAAS, F. L., 1959: Mutation induction and macromolecular synthesis in bacteria. – P. N. A. S. (Wash.) **45**, 709–722.

[15] DOTY, R., MARMUR, Y., and SUEOKA, N., 1959: Heterogeneity in deoxyribonucleic acids. – Nature 183, 1427–1431.

[16] FINCHAM, J. R. S., 1959: The role of chromosome loci in enzyme formation. – Proc. X. Internat. Cong. Gen., pp. 355–363. McGill Univ. Press, Montreal.

[17] FRAENKEL-CONRAT, H., SINGER, BEATRICE A., and WILLIAMS, R. C., 1957: The nature of the progeny of virus reconstituted from protein and nucleic acid of different strains of Tobacco Mosaic Virus. – In "The Chemical Basis of Heredity", pp. 501–541. Johns Hopkins Press, Baltimore.

[18] FREESE, E., 1957: The correlation effect for a histidine locus of *Neurospora crassa*. – Genetics **42**, 671–684.

[19] FREESE, E., 1958: The arrangement of DNA in the chromosome. – Cold Spring Harbor Symp. Quant. Biol. **23**, 13–18.

[20*a*] FREESE, E., 1959*a*: The difference between spontaneous and base-analogue induced mutations of phage T4. – P. N. A. S. (Wash.) **45**, 622–633.

[20*b*] FREESE, E., 1959*b*: The specific mutagenic effect of base analogous on phage T 4 – J. Molecular Biol. **1**, 87–105.

[21] GILES, N. H., 1959: Mutations at specific loci in *Neurospora*. – Proc. X. Internat. Cong. Gen., pp. 261–279. McGill Univ. Press, Montreal.

[22] GOLDSCHMIDT, R., 1955: Theoretical Genetics. – Univ. Calif. Press, Berkeley.

[23] GOLOMB, S. W., WELCH, L. R., and DELBRÜCK, M., 1958: Construction and properties of comma-free codes. – Biol. Medd. Dan. Vid. Selsk. **23**, no. 9, 1–32.

[24] HARTMAN, P. E., 1957: Transduction: a comparative review. – In "The Chemical Basis of Heredity", pp. 408–462. Johns Hopkins Press, Baltimore.

[25] HERSHEY, A. D., and CHASE, M., 1952: Independent function of viral protein and nucleic acid in growth of bacteriophage. – J. Gen. Physiol. **36**, 39–56.

[26] HOROWITZ, N. H., FLING, M., MACLEOD, H. L., and SUEOKA, N., 1959: Multigenic control of tyrosinase synthesis in *Neurospora*. (Abstract.) – Genetics, in press.

[27] HUNT, J. A., and INGRAM, V. H., 1958: The chemical effects of gene mutations in some abnormal human hemoglobins. — In "Symp. on Protein Structure" ed. by A. NEUBERGER. Methuen, London; Wiley, New York.

[28] JACOB, F., 1958: Genetic control of viral functions. — Harwey Lecture; sept. 1958. Thomas. Springfield, Ill. (In press).

[29] LEDERBERG, J., 1950: Isolation and characterization of biochemical mutants of bacteria. — In "Methods in Medical Res.", Vol. 3, pp. 5–22. The Year Book Publs., Chicago.

[30] LEDERBERG, J., 1955: Recombination mechanisms in bacteria. — J. Cell. Comp. Physiol. 45, (Suppl. 2), pp. 75–107.

[31] KELLENBERGER, E., RYTER, A., and SÉCHAUD, Y., 1959: Electron microscope study of DNA-containing plasmas. — Journ. Biophys. Biochem. Cyt. (In press).

[32] LITMAN, ROSE M., and PARDEE, A. B., 1956: Production of bacteriophage mutants by a disturbance of deoxyribonucleic acid metabolism. — Nature 178, 529–530.

[33] LITMAN, ROSE M., and PARDEE, A. B., 1959: Mutations of bacteriophage T2 induced by bromouracil in the presence of chloramphenicol. — Virology 8, 124–127.

[34] LURIA, S. E. FRASER, DOROTHY K., ADAMS, JANE N., and BURROWS, JEANNE W., 1958: Lysogenization, transduction and genetic recombination in Bacteria. — Cold Spring Harbor Symp. Quant. Biol. 23, 71–82.

[35] MAALØE, O., 1959: Synchronous growth. — In "The Bacteria" II, chapter 3. In press, Academic Press, New York.

[36] MESELSON, M., and STAHL, F. W., 1958: The replication of DNA. — Cold Spring Harbor Symp. Quant. Biol. 23, 9–12.

[37] PARDEE, A. B., 1959: Mechanisms of control of enzyme synthesis and enzyme activity in bacteria. — Ciba Found. Symp. on "Regulation of Cell Metabolism", pp. 295–304. Churchill, London.

[38] PONTECORVO, G., 1958: Trends in Genetic Analysis. — Columbia Univ. Press, New York.

[39] PRITCHARD, R. H., 1955: The linear arrangement of a series of alleles of *Aspergillus nidulans*. — Heredity 9, 343–371.

[40] REISSIG, J. K., 1959: Forward and backmutation at the *Pyr*$_3$ locus of *Neurospora*. (Abstract). — Heredity 13, 144.

[41] ROMAN, H., and JACOB, F., 1958: A comparison of spontaneous and ultraviolet-induced allelic recombination with reference to the recombination of outside markers. — Cold Spring Harbor Symp. Quant. Biol. 23, 155–160.

[42] SINGER, S. J., and ITANO, H. A., 1959: On the asymmetrical dissociation of human hemoglobin. — P. N. A. S. (Wash.) 45, 174–183.

[43] SPOONER, E. T. C., and STOCKER, B. A. D. (Editors), 1956: "Bacterial Anatomy." — 6. Symp. Soc. Gen. Microbiology. Cambridge Univ. Press.

[44] STARLINGER, P., and KAUDEWITZ, F., 1956: Ein pseudoallel-spezifischer Suppressor bei *Salmonella typhimurium*. — Z. f. Naturforsch. 11, 317–329.

[45] TAYLOR, J. H., 1957: The time and mode of duplication of chromosomes. — Amer. Nat. 91, 209–221.

[46] TAYLOR, J. H.: 1959: The organization and duplication of genetic material. — Proc. X. Internat. Cong. Gen., pp. 63–78. Mc Gill Univ. Press, Montreal.

[47] TIMOFÉEFF-RESSOVSKY, N. W., 1934: The experimental induction of mutations. — Biol. Rev. (Cambridge), 9, 411–432.

[48] WATSON, J. D., 1957: X-ray studies on RNA and the synthetic polyribonucleotides. — In "The Chemical Basis of Heredity", pp. 552–556. Johns Hopkins Press, Baltimore.

[49] WESTERGAARD, M., 1957: Chemical mutagenesis in relation to the concept of the gene. — Experientia 13, 224–234.

[50] WINKLER, H., 1935: Chimären und Burdonen. Die Lösung des Propfbastardproblems. — Der Biologe 9, 279–290.

[51] WITKIN, EVELYN M., 1959: Post irradiation metabolism and the timing of ultraviolet-induced mutations in Bacteria. — Proc. X. Internat. Cong. Gen., pp. 280–299. McGill Univ. Press, Montreal.

[52] WOODWARD, D. O., 1959: Enzyme complementation in vitro between adenylosuccinase-less mutants of *Neurospora crassa*. — P. N. A. S. (Wash.) 45, 846–850.

[53] WOODWARD, V. W., DEZEEUW, J. R., and SRB, A. M., 1954: The separation and isolation of particular biochemical mutants of *Neurospora* by differential germination of conidia, followed by filtration and selective plating. — P. N. A. S. (Wash.) 40, 192–200.

[54] WOLLMAN, E., JACOB, F., and HAYES, W., 1956: Conjugation and genetic recombination in *Escherichia coli*, K 12. — Cold Spring Harbor Symp. Quant. Biol. 21, 141–162.

Some implications of a new interpretation of chromatid aberrations induced by ionizing radiations and chemical agents

S. H. REVELL

Chester Beatty Research Institute, Institute of Cancer Research:
Royal Cancer Hospital, London, S.W. 3.

An interpretation of chromatid aberrations induced by ionizing radiations and chemical agents has already been published (REVELL 1955, 1958, 1959), which differs significantly from that generally accepted: chromatid and isochromatid discontinuities, which it is customary to regard as surviving single or double primary breakages, are interpreted instead as arising from minute intrachanges which unravel during chromosome contraction to metaphase.

It has been pointed out (1959) that acceptance of this hypothesis would make it unlikely that the **primary event**, occurring just as the chromosome is traversed by the ionizing particle, is an actual chromatid breakage; and also, therefore, unlikely that the conjugation of a pair of such primary events (defined as **exchange initiation**) consists of a reciprocal chromatid cross-union. Thus, the temporary capacity of an affected chromosome to conjugate with another in an exchange initiation does not consist in a chromatid break remaining open, but, rather, must be some labile tendency to undergo a point-pairing with another site similarly affected.

The object of this short paper is to summarize certain other implications of the exchange interpretation.

1. As already pointed out (1959), if the two early stages in exchange induction, i. e. those which are separable by manipulation of the X- or γ-ray dose, are not chromatid breakage and reunion, then it follows that the chromatid exchange itself is structurally established at some later stage. The state of exchange initiation is thus a pre-mutational condition in respect of the subsequent exchange. From evidence presently available it does not seem posssible to decide whether or not an exchange initiation is irreversibly committed to exchange realisation.

2. In root meristem cells of *Vicia faba* the stage just before the beginning of prophase is the latest one at which X-rays can induce chromatid exchanges to appear at the following metaphase (1959). It is therefore possible (though not necessary for the validity of the exchange interpretation) that the exchange process which must be envisaged is developmentally homologous with the meiotic one.

3. It has always been difficult to see how single ionizing particles, or chemical agents at very low concentration, could actually break chromosomes. It seems easier to see how they

might initiate exchange, the exchange itself being a subsequent chromosome development possibly akin to meiotic exchange. Thus the chromosomes in the nuclear environment could be considered to perform the work of the exchange, the chemical or radiation only predisposing them to do so. According to this conception, both primary event and exchange initiation could be of very small magnitude in terms of chemical change, though subject to amplification by subsequent chromosome development.

4. If the exchange interpretation is correct then the spatial arrangement of chromosomes in the resting nucleus must be an important factor affecting sensitivity to the induction of all types of chromatid aberration, whereas according to the breakage-and-reunion interpretation the component of sensitivity due to simple breakage (and consequent deletion) is unaffected by spatial considerations. The exchange interpretation thus appears to offer greater scope in accounting for large differences in sensitivity, so far as the induction of chromatid aberrations is concerned. It seems less difficult to conceive of nuclei differing greatly in their capacity to develop exchanges in response to chemical or radiation treatment than it is to envisage large inherent differences in chromosome breakability.

5. During the stage when chromatid changes are initiated – the stage usually called "post-split" – the chromosomes need not be differentiated into chromatids: it is not necessary to assume that the primary events take place in the chromatids which are finally observed to be exchanged.

Acknowledgements

This work has been supported by grants to the Chester Beatty Research Institute (Institute of Cancer Research: Royal Cancer Hospital) from the British Empire Cancer Campaign, the Jane Coffin Childs Memorial Fund for Medical Research, the Anna Fuller Fund, and the National Cancer Institute, U. S. Public Health Service.

References

REVELL, S. H., 1955: A new hypothesis for chromatid changes. In Proceedings of the Radiobiology Symposium, Liège 1954 pp. 243–253. London: Butterworth.

REVELL, S. H.1958: A new hypothesis for the interpretation of chromatid aberrations, and its relevance to theories for the mode of action of chemical agents. Ann. N.Y. Acad. Sci., **68**, 802–807.

REVELL, S. H., 1959: The accurate estimation of chromatid breakage, and its relevance to a new interpretation of chromatid aberrations induced by ionizing radiations. Proc. Roy. Soc. B, **150**, 563–589.

Chromatid aberrations induced by ethyleneimines

C. H. Ockey

Department of Cytogenetics, Christie Hospital and Holt Radium Institute, Manchester 20

Introduction

In a previous study in which root-tips of *Vicia faba* were treated for one hour with a range of concentrations of acetylethyleneimine (AEI) or triethylene melamine (TEM) (Ockey, 1957), it was found that the proportions of the different aberration types observed at a fixed 24 hour recovery period after treatment changed in a regular manner. The frequency with which the different chromosomes were attacked by the agent also changed with concentration. However it could not be decided from the data available whether these changes were directly due to increase in concentration or whether a time factor had been introduced by the different degrees of mitotic inhibition produced by the different concentrations.

To investigate this problem further experiments were conducted in which a single concentration of an alkylating agent was used and a wide range of recovery periods sampled. One such experiment using freshly prepared triethylene melamine at a concentration of 1×10^{-4} molar will be described here. As in the case with other alkylating agents, the aberrations induced by ethyleneimines are localised in or near the heterochromatic regions of *Vicia* chromosomes. It was decided to investigate more closely both the nature and distribution of heterochromatin and its relationship to the aberrations induced by TEM.

Techniques

Seedlings of the variety Seville Long Pod were grown in individual tubes at 19° C and the lateral roots treated. The water in the culture tubes was replaced by a freshly prepared aqueous solution of TEM at 1×10^{-4} molar for 1 hour; the roots were then thoroughly washed and allowed to recover in tap water at 19° C. For the last four hours of the selected recovery periods roots were immersed in 0.03 % colchicine to accumulate metaphases. They were then fixed in La Cour's 2 BD, bleached, hydrolysed and stained in leuco basic fuchsin in the usual way. Slides were prepared from three seedlings at each recovery period.

As many features of the results were more readily accounted for on the basis of Revells (1952) interpretation of chromatid exchanges rather than on the orthodox 'breakage and reunion' interpretation, his method of analysis was adopted.

100 cells with aberrations were scored from each seedling and the mean value of the two or three seedlings scored at each recovery period calculated. The results are expressed as exchanges per 100 aberrant cells and shown graphically together with the standard errors of several aberration types (Fig. 1). At some of the earliest recovery periods fewer than 100 cells with aberrations could be scored in each seedling.

Results

In common with other alkylating agents aberrations were rarely observed until 10 hours after treatment. At the 12 hour recovery period about $10\,^0/_0$ of the cells showed aberrations and at this time there was a high frequency of intrachanges Type 4 (sister union) while the number of interchanges was then low. With increase in recovery time, the number of intrachanges Type 4 steadily declined and that of interchanges rose steeply (Fig. 1).

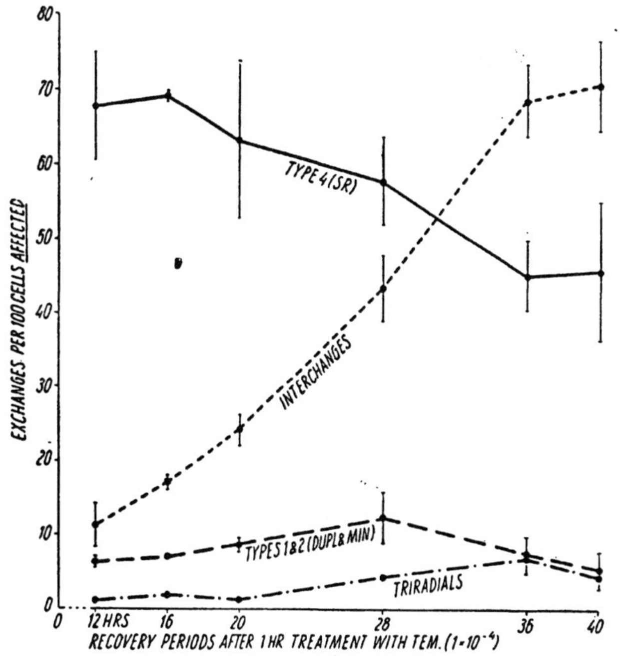

Fig. 1. Change in proportion of the different types of aberration observed at different recovery periods after a 1 hour treatment with TEM 1×10^{-4} molar. Standard errors are represented as vertical lines for some aberration types.

The number of triradials increased with increase in recovery period up to 36 hours, but they were always less frequent than the interchanges and their rate of increase was also lower. Although interchange number continued to increase after 36 hours the number of triradials declined. This would be expected on the exchange hypothesis, according to which a triradial is derived from an interchange combined with one Type 4 intrachange. The decline in Type 4 at the latter times now controls the frequency of triradials, whereas previously this was principally controlled by the lower frequency of interchanges.

On the exchange hypothesis, each of the three detectable types of intrachange, Types 1 (duplication), 2 (minute) and 4 (sister union) would be expected to occur in equal numbers. However Types 1 and 2 can only be scored when the loop involved in their formation is above a certain minimum size whereas the Type 4 intrachange will always yield a detectable result. Consequently the Type 4 intrachanges will outnumber the total of Types 1 and 2.

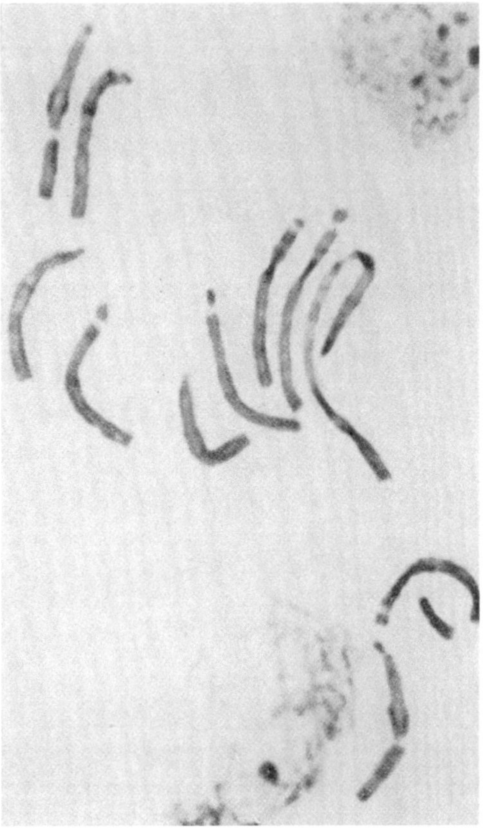

Fig. 2. Part of cell from root subjected to 14 days at 5°C, followed by pretreatment with hydroxyquinoline for 2 hours prior to fixation, showing positive and negative heterochromatin. × 2,100.

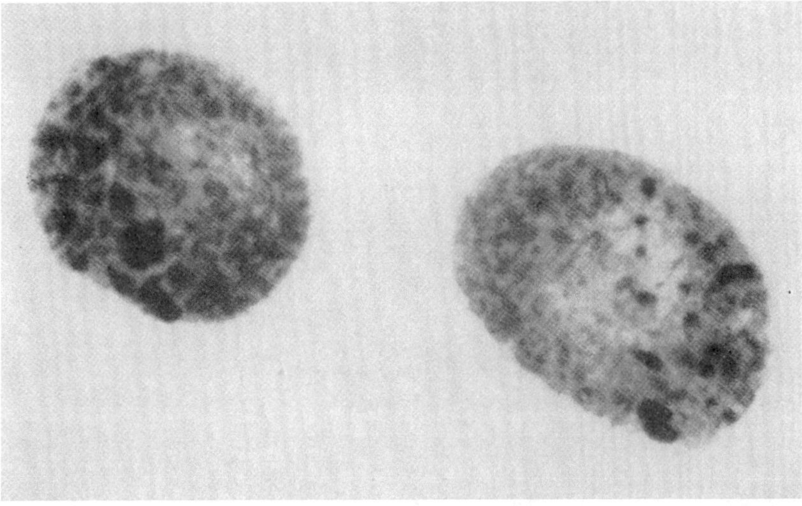

Fig. 4. Two cells in very early interphase. Right hand cell shows 20 chromocentres. The two largest (those in chromosome 5) show homologous association. × 2,100.

In this experiment there were 7 times as many Type 4 intrachanges as Types 1 and 2. It might also be expected that the frequency curve for Types 1 and 2 intrachange would parallel that of the Type 4 with time. However from 12 to 28 hours there was a slight increase in the number of Type 1 and 2 intrachanges while the number of Type 4 was decreasing, but after this period the two frequency curves did decrease in parallel (Fig. 1). The loops involved in the formation of the intrachanges during this early period were very small and many of the Type 1 and 2 intrachanges could not be detected; the difference in the curves at this time compared with the later recovery periods is therefore not due to differences in the mechanism of induction between Type 1 and 2 and Type 4, but to differences in loop size in their formation.

At the early recovery periods Type 2 intrachanges outnumbered Type 1 by a factor of about 4, whereas at later periods this was reduced to one of 1.5. This again probably arises from the greater ease of detecting a very small minute than a duplication of similar size. These differences found in the frequencies of the different intrachange types with recovery time can be accounted for satisfactorily if we assume that there is a decrease in loop size during interphase, the loops being largest at the 36 hour recovery period which corresponds to the earliest part of the period sensitive to TEM, and decreasing in size as interphase proceeds.

Heitz in 1932 indicated that the number of chromocentres in resting nuclei of *Vicia faba* varied from 8 to 20 and that these chromocentres corresponded to regions of telophase chromosomes which retained their staining ability throughout interphase. Large heterochromatic regions occurred in the middle of the long arm of the shorter S chromosomes and on either side of the centromere of the long L chromosome. La Cour (1951) and McLeish (1953) found that cold treatment of roots led to the appearance of heterochromatic regions as non-or pale-staining segments at metaphase, and McLeish distinguished eight such regions in the haploid complement. Tjio and Levan (1950) using oxyquinoline distinguished dark staining regions in prophase chromosomes from roots grown at normal temperatures located in the median positions as described by Heitz. By pretreating roots grown at 5° C for 14 days with 0.002 molar hydroxyquinoline both types of region could be distinguished (Fig. 2). These dark and pale staining regions have for convenience been termed positive and negative heterochromatin respectively.

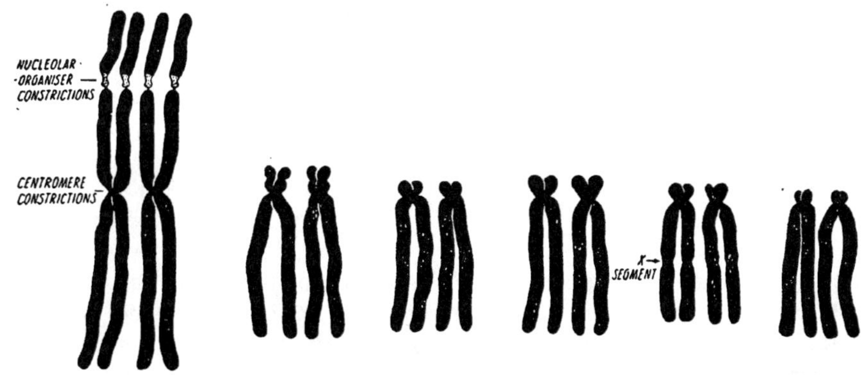

Fig. 3. Idiogram of the six pairs of chromosomes in *Vicia faba*. Note the X segment in the shortest chromosome (pair 5).

As it was possible to identifiy all six pairs of chromosomes in the complement of *Vicia faba* in most cells (Fig. 3), the position of these positive and negative regions could be plotted accurately from observations on many metaphases from cold treated root-tips. The position of the darker positively heterochromatic segments was found to be related to that of the pale negative regions, the former generally being located on the proximal side of the latter and not infrequently at some distance from these, as with the largest regions in chromosome 5 (Fig. 2). Expression of one type of segment was correlated with that of the other. This was particularly clear in chromosome 5, in which, if the positively heterochromatic region was large, the negative one was also well defined, the rest of the chromosome being unaffected. The number and position of the positively heterochromatic segments indicated that these probably correspond to the interphase chromocentres (Fig. 4), the large pair of chromocentres being the median segments of chromosome 5. It can also be seen from Fig. 4 that the parallel arrangement of the chromosomes found at telophase is retained for at least part of interphase. The effect of cold treatment suggests that the positively heterochromatic regions may be tightly coiled at the expense of the negative ones. In roots grown at normal temperature this difference in coiling is less marked.

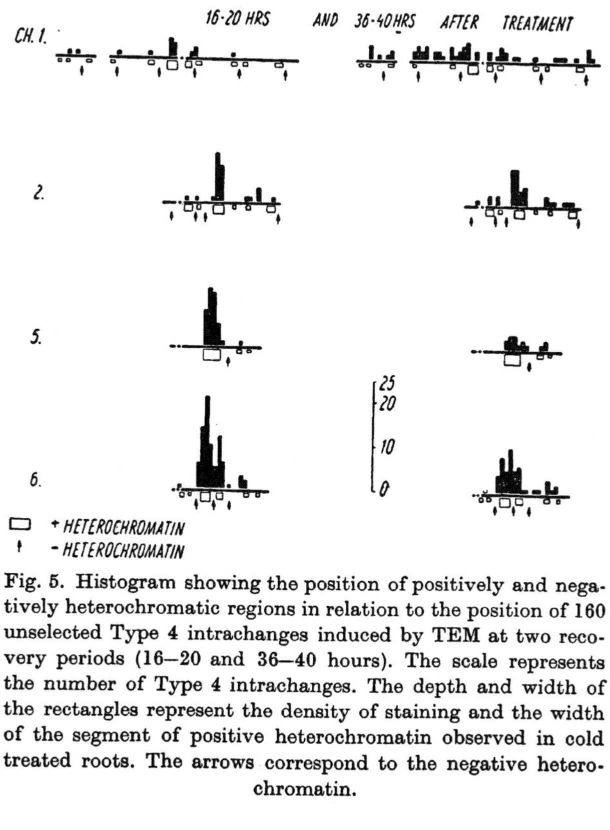

Fig. 5. Histogram showing the position of positively and negatively heterochromatic regions in relation to the position of 160 unselected Type 4 intrachanges induced by TEM at two recovery periods (16—20 and 36—40 hours). The scale represents the number of Type 4 intrachanges. The depth and width of the rectangles represent the density of staining and the width of the segment of positive heterochromatin observed in cold treated roots. The arrows correspond to the negative heterochromatin.

Fig. 5 shows the positions of these regions for chromosomes 1, 2, 5 and 6. Chromosomes 3 and 4 possessed small regions of positive heterochromatin in a median position. Also shown in Fig. 5 are the positions of 160 unselected Type 4 intrachanges from seedlings at early and late recovery periods. Chromosome 3 and 4 are omitted as they are involved in aberrations less frequently than chromosomes 1, 2, 5 and 6.

It is clear from the figure that exchanges occur predominantly in the positively heterochromatic segments and are infrequent in the negative ones. This is most obvious in chromosome 5 where the two regions are clearly separated. Up to 20 hours the degree of localisation is very high and chromosome 1 participates in intrachanges infrequently compared with the other chromosomes. With increase in recovery time, the degree of localisation becomes less and less marked and at 36 to 40 hours, chromosome 5 is involved much less frequently than it was at earlier periods while there is no such striking change with chromosomes 2 and 6. Chromosome 1 is far more frequently involved at the later periods.

Fig. 6 shows, for each of three different recovery periods, the position of 70 interchanges between the same 4 chromosomes. Where the interchange positions are crowded together the area has been shaded and the number involved indicated. Localisation in the positive heterochromatin is again apparent particularly at the erarliest recovery times of 12 to 20 hours with 50% of the interchanges localised in the median segments of chromosomes

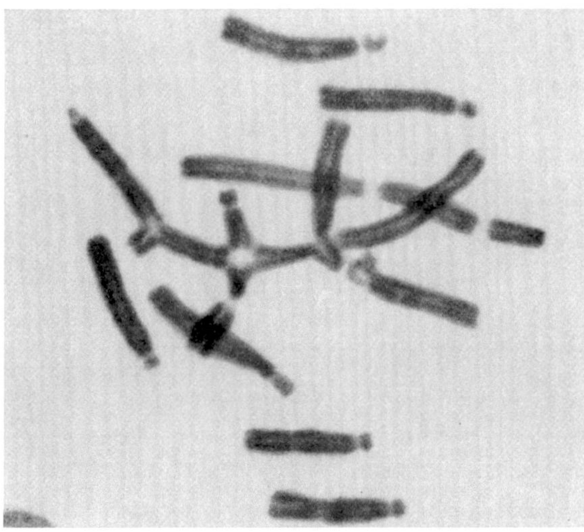

Fig. 7. Multiple interchange involving chromosomes 1, 3, 4 and pair 6. Each pair of interchange points located at a different distance from the centromere.

2, 5 and 6. At 36 hours this localisation becomes less marked with only 13.6 % of the interchanges in these heterochromatic regions. Chromosome 5 now participated less frequently in interchanges while chromosome 1 was involved more than at the early recovery periods, a similar situation to that found for the Type 4 intrachanges.

The proportion of interchanges that are homologous also decreases with increase in recovery time, 22 of 70 being homologous at 12–20 hours but only 12 at 36 hours.

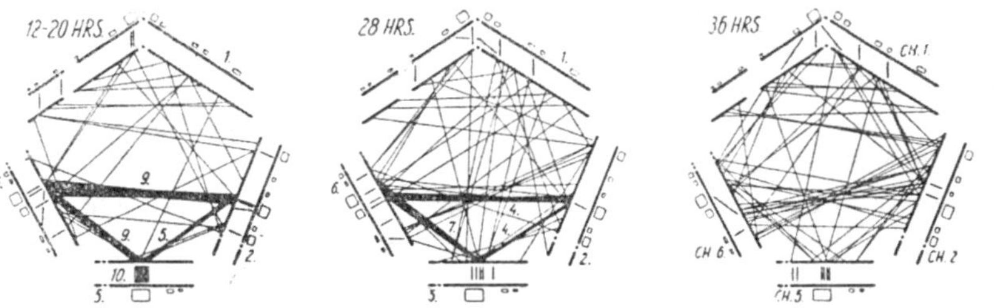

Fig. 6. Position of 70 interchanges between chromosomes 1, 2, 5 and 6 at 12–20, 28 and 36 hours after treatment for 1 hour with TEM at 1×10^{-4} molar.

It is also clear that most interchanges are between loci which are equidistant from the centromeres of the chromosomes involved and in many of the multiple interchanges found, each pair of interchange points was located at a different distance from the centromere (Fig. 7).

The complex aberrations produced by TEM could always be interpreted as a combination of the basic exchange types defined by REVELL and in all cases one chromatid only was involved at a single chromosome locus.

Discussion

The changes in frequency and position of exchanges described are those anticipated on the exchange hypothesis, which requires contact between chromosome parts for initiation of the exchange. Increase in nuclear size during interphase and less uncoiling in the heterochromatic segments are both important in determining the type of aberration induced (Fig. 8). Early in interphase, corresponding to the late recovery periods, the nucleus is small and the chromosomes parallel to one another; at this stage contact between different chromosomes is more likely than in a bigger nucleus and one would expect interchanges to form a higher proportion of the total aberrations than with treatment later in interphase. This agrees with the

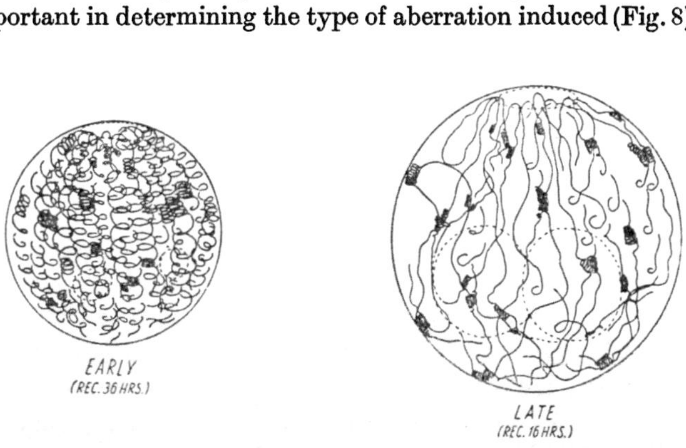

Fig. 8. Diagram showing postulated nuclear structure at early and late periods in the part of interphase sensitive to the action of TEM.

observations and provides an explanation for the greater number of aberrations found at this stage.

As the nucleus increases in size there is less contact between chromosomes and intrachanges form a steadily increasing proportion of the total aberrations. As the chromosomes become more diffuse (assuming uncoiling extends into part of interphase), there will be a greater contrast between the degree of coiling of heterochromatic and euchromatic regions and contact within the loops of the chromosome will be more likely in the heterochromatin. With the increase in nuclear volume a decrease in the size of the loops seems a likely event to account for the changes in the frequency of Type 1, 2 and 4 intrachanges with time. This decrease in loop size could well be a factor determining the greater degree of localisation in cells treated later in interphase.

This increase in localisation is also one reason for the change in the ratio in which S and L chromosomes are affected. The S chromosomes contain more positive heterochromatin relative to their length than do the L and they become increasingly affected as treatment is given later in interphase.

The change in the frequency of homologous to heterologous interchanges with time also follows from increasing localisation in heterochromatin combined with a greater chance of fusion of homologous chromocentres because they lie at the same distance from the pole of the nucleus.

Although the explanation of the tendency for aberrations to be localised in the heterochromatin is essentially a mechanical one to do with different degrees of coiling, also involved is the well known tendency for fusion of chromocentres which will promote contact between chromosomes and within chromosome loops. The unexpected behaviour of the median segment of chromosome 5 compared with those of chromosomes 2 and 6 (which was also found in other experiments) suggests however that some physiological changes may also play a part in deciding the regions affected. This is likely since it cannot be considered that the insensitivity of the latter part of interphase to alkylating agents is due to loss of contact between chromosome parts as chromocentres are present throughout interphase. The loss of sensitivity must rather be bound up with the occurrence of DNA synthesis and chromosome duplication, events which appear to demarcate the sensitive and insensitive parts of interphase.

It is evident from these results that the alkylating agent has a direct effect on some chromosome regions in contact at the time of treatment, probably by a process of alkylation of chromosome components. The chromosome regions affected must subsequently be held together, perhaps by some process involving fusion of nucleoprotein threads or by exchange of two strands, one from each chromosome; such a process would be compatible with Taylor's (1958) two-stranded model for a chromosome prior to DNA synthesis. It is unlikely however that the aberrations are actually produced at this time, before the chromosome has duplicated. This duplication cannot occur before DNA synthesis (Taylor, 1958). The aberrations which are uniformly of the chromatid type must be produced at some period between DNA synthesis and prophase, perhaps at the time of chromosome duplication. In terms of Taylor's two-stranded model this process could involve the synthesis of one of the new DNA strands on the pattern of the strands exchanged at the time of treatment.

The change in the proportion of different aberration types observed in the recovery period experiments are of a similar type to the changes described for the concentration range experiments (Ockey, 1957). In the concentration range experiments, the increase in mitotic inhibition produced with the increase in concentration introduced a time factor and resulted in the change in proportion of the different aberration types.

Summary

1. At 12 to 16 hours after treatment of *Vicia faba* root-tips with 1×10^{-4} molar TEM for 1 hour there is a high proportion of intrachanges Type 4 (sister union) while that of interchanges is low. With increase in recovery time the number of intrachanges Type 4 decrease and that of interchanges increase.

2. In cold treated roots, two types of heterochromatic regions have been described, one pale staining and termed negative heterochromatin corresponds to the 'starved' regions described by La Cour (1951), the other staining darkly termed positive heterochromatin corresponds to the chromocentres of the interphase nucleus.

3. At 12–16 hours after treatment aberrations were localised in the positively heterochromatic segments but became less localised with increase in recovery time. Chromosome 5 was involved far more frequently at the early periods than later, while chromosome 1 was involved more at the later times.

4. These changes with time are interpreted in terms of increase in nuclear size, less uncoiling in positive heterochromatin than euchromatin and variation in contact between and within chromosomes during interphase.

References

Heitz, E., 1932: Die Herkunft der Chromocentren. Planta: 18, 571–636.

La Cour, L. F., 1951: Heterochromatin and the organisation of nucleoli in plants. Heredity: 5, 37–50

McLeish, J., 1953: The action of maleic hydrazide in *Vicia*. Heredity (suppl. vol.): 6, 125–148.

Ockey, C. H., 1957: A quantitative comparison between the cytotoxic effects produced by proflavine, acetylethyleneimine, and triethylene melamine on root-tips of *Vicia faba*. J. Genet: 55, 525–549.

Revell, S. H., 1952: A new interpretation of the chemical induction of chromosomal aberrations. Ann. Rep. Brit. Emp. Cancer Camp.: 30, 42–55.

Taylor, J.H., 1958: Sister chromatid exchanges in tritium labelled chromosomes. Genetics 44, 515–529.

Tjio, J., and Levan, A., 1950: The use of oxyquinoline in chromosome analysis. Anal. Est. Exp. Aula. Dei: 2, 21–64.

Über die radiomimetische Wirkung von Äthylalkohol bei *Vicia faba*

R. Rieger und A. Michaelis

Institut für Kulturpflanzenforschung der Deutschen Akademie der Wissenschaften, Gatersleben, Kr. Aschersleben, und Institut für Genetik der Martin-Luther-Universität Halle-Wittenberg

1. Einleitung und Methodik

Der Äthylalkohol interessiert als radiomimetisches Agens, weil es sich hier um eine sehr einfache, nicht-dissoziierende organische Verbindung handelt, die überdies unter Bedingungen partieller oder vollständiger Anaerobiose in erheblichen Mengen im pflanzlichen Stoffwechsel gebildet wird, womit eine Beziehung zur stofflichen Grundlage der spontanen Mutabilität hergestellt ist.

Ausgangspunkt unserer Untersuchungen war die Feststellung, daß die Quellung von *Vicia*-Samen unter partiell anaeroben Bedingungen (Einquellung in stehendem Wasser, in Stickstoff- oder CO_2-Atmosphäre) zur Produktion automutagener Substanzen führt (Rieger und Michaelis 1958, 1959), die in der Lage sind, in erheblichem Umfang Chromosomenaberrationen zu induzieren. Durch Destillation ließ sich das verantwortliche Agens aus den Samen bzw. dem Einquellwasser isolieren, stark anreichern und als Äthylalkohol identifizieren (Michaelis, Ramshorn und Rieger 1959), der damit eindeutig als automutagene Substanz ausgewiesen ist. Es lag nahe, den Versuch zu unternehmen, nähere Aufschlüsse über sein Wirkungsvermögen und seine Wirkungsweise zu erhalten.

Alle im folgenden zu schildernden Versuche liefen, wenn nicht ausdrücklich anders angegeben, in drei Wiederholungen. Pro Versuchsstufe und Wiederholung wurden 250 Zellen aus je fünf Präparaten ausgewertet, d. h. 750 Zellen in drei Wiederholungen. Behandelt wurden jeweils 20 Primärwurzeln auf einem 50 ml Gefäß im Thermostaten bei 24° C. Die gleiche Temperatur wurde während der Erholungszeiten auf fließendem Leitungswasser beibehalten. Die Präparation erfolgte nach der Feulgen-Quetschmethodik im Anschluß an eine zweistündige Vorbehandlung mit 0,05 %iger Colchizinlösung. Alle Präparate wurden über Trockeneis-Alkohol-Euparal haltbar gemacht.

2. Ergebnisse

Äthylalkohol löste außer Isolocusbrüchen und reziproken Chromatidentranslokationen einen geringeren Anteil von Duplikations-Deletionen, Deletionen und Triradialen und wenige nicht an Reunionen beteiligte Chromatidenbrüche aus. Die induzierten Aberrationen waren ausschließlich auf chromatidaler Basis erfolgt. Die Bruchverteilung war präferentiell und entsprach bei bevorzugter Beteiligung der heterochromatischen Segmente und der fünf Paare kleiner Chromosomen den im Zusammenhang mit der Samenquellung unter partieller Anaerobiose erhaltenen Befunden (Abb. 1).

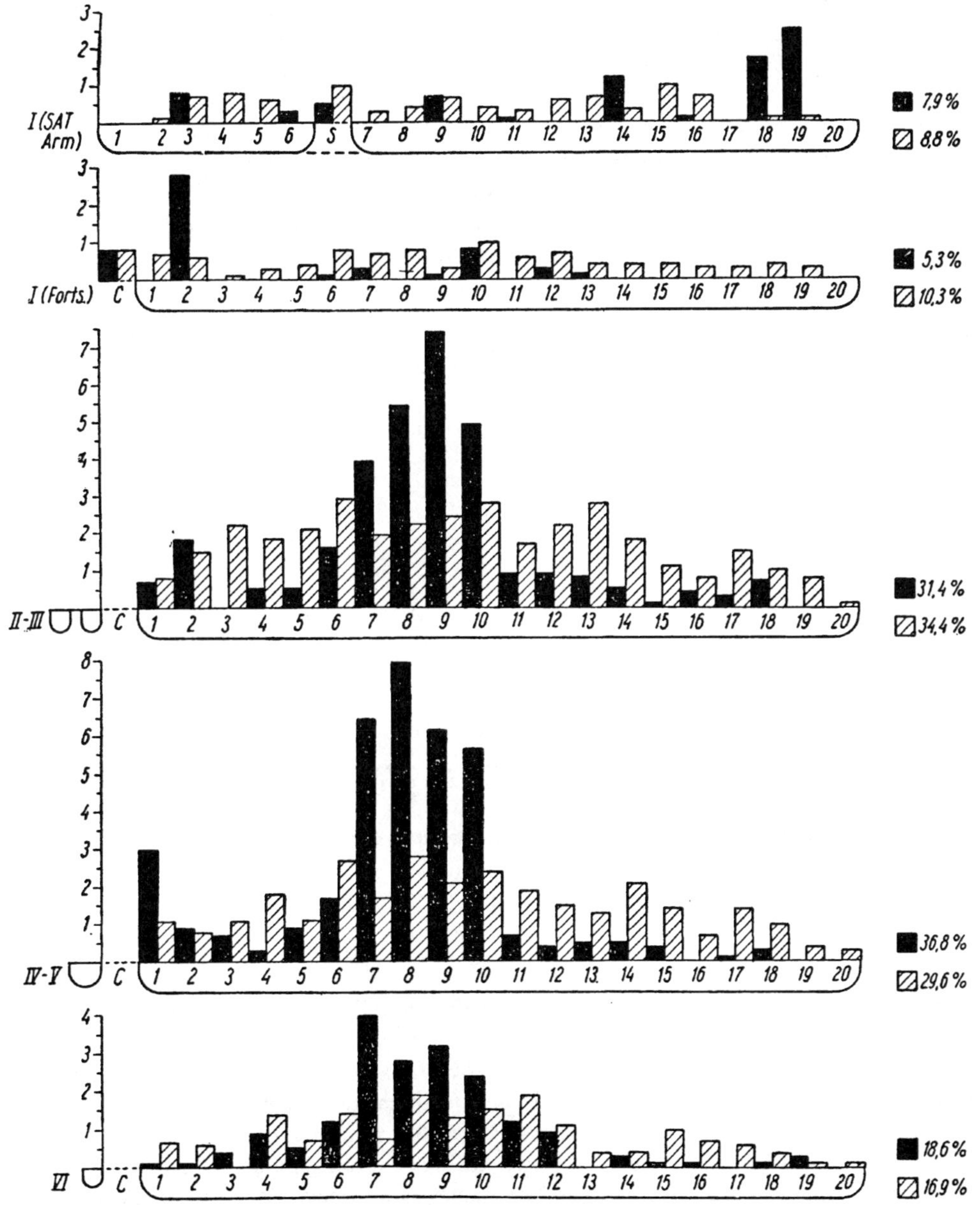

Abb. 1. Die prozentuale Beteiligung der Einzelsegmente des in 4 Chromosomengruppen unterteilten Chromosomensatzes (SAT-Chromosom in 2 Hälften untereinander) an der gesamten erfaßten Aberrationsfrequenz nach Einwirkung von Äthylalkohol (schwarze Säulen) und Röntgenstrahlen (schraffierte Säulen). Diploide Zellen besitzen 2 der Gruppe I, 4 der Gruppe II–III, 4 der Gruppe IV–V und 2 der Gruppe VI zugehörige Chromosomen. Die Prozentangaben rechts beziehen sich auf die Beteiligung der jeweiligen Gruppe am Gesamtaberrationsgeschehen.

Versuch 1: Einwirkung steigender Alkoholkonzentrationen

Die Behandlungszeit der Wurzeln bei 22° C betrug 24 Stunden. Anschließend folgte eine 24stündige Erholungszeit, an die sich nach der Colchizinbehandlung die Fixierungen anschlossen. 1 molare Äthylalkoholkonzentrationen erwiesen sich bei 24stündiger Einwirkung als letal, Konzentrationen unter 10^{-3} Mol blieben wirkungslos. Von 10^{-3} bis 5×10^{-1} erhöhte sich der Anteil von Zellen mit Chromosomenaberrationen in der in Abb. 2a dargestellten Weise. Ein dem Anstieg der einwirkenden Konzentrationen proportionaler Anstieg in der Zahl aberranter Zellen liegt nicht vor, so daß sich keine rein quantitative Beziehung zwischen applizierter Dosis und ausgelöstem Effekt ergibt. Erstaunlich hoch ist gemessen an der Anzahl der Isolocusbrüche der Anteil an reziproken Chromatidentranslokationen, die den Isolocusbrüchen sehr dicht folgen (Abb. 2b). In dieser Hinsicht scheint sich der Alkohol von anderen radiomimetischen Agenzien zu unterscheiden, da nach deren Einwirkung meist sehr

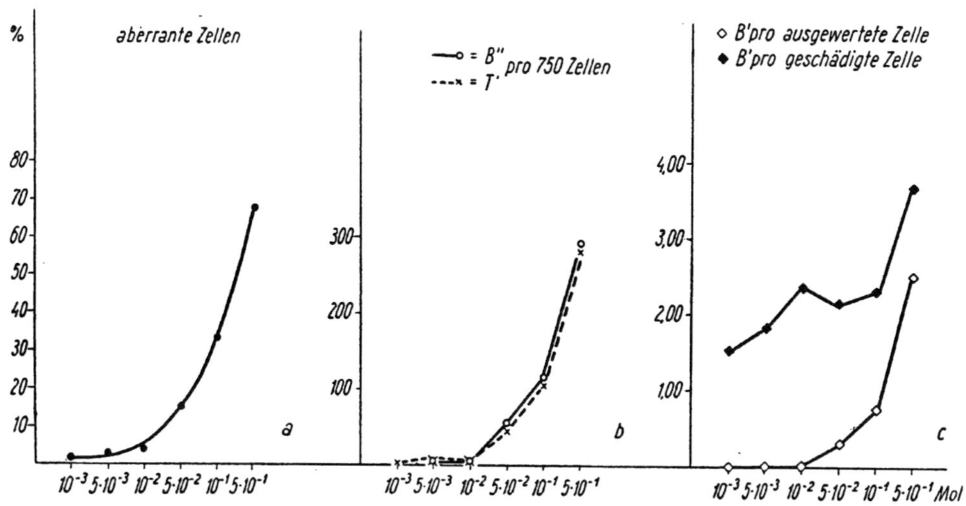

Abb. 2. a—c. Der Einfluß verschiedener Äthylalkoholkonzentrationen auf die Primärwurzeln.
a) Der prozentuale Anteil geschädigter Zellen,
b) Die absolute Anzahl von Isolocusbrüchen (B'') und Chromatidentranslokationen (T'),
c) Die Anzahl an Chromatidenbrüchen pro ausgewertete und pro geschädigte Zelle.

viel mehr Isolocusbrüche als Translokationen auftreten. Ganz rechts in der Abb. 2 ist die Anzahl der Chromatidenbrüche pro geschädigte und ausgewertete Zelle in Beziehung zur Dosis gesetzt worden. Für jeden Isolocusbruch, jede Translokation und Deletion wurden zwei, für Duplikations-Deletionen und Triradiale drei Chromatidenbrüche in Rechnung gestellt. Während im Bereich der Konzentrationen 10^{-2} bis 10^{-1} Mol kein Anstieg der Chromatidenbrüche pro geschädigte Zelle erfolgte, trat eine starke Erhöhung bei Anwendung von 5×10^{-1} Mol Alkohol ein. Abgesehen von den niedrigsten Konzentrationen erhöhte sich die Zahl der Chromatidenbrüche pro ausgewertete Zelle kontinuierlich auf Grund der zunehmenden Anzahl von geschädigten Zellen.

In Abb. 3 haben wir den Versuch unternommen, an Hand der verwendeten Konzentrationen und der Anzahl ausgelöster Isolocusbrüche und Translokationen eine Dosis-Effektkurve aufzustellen. Derartige Versuche sind im Falle chemischer Mutationsauslösung mit Vorbehalten belastet, da die effektive Dosis des Agens, die die Aberrationen induziert, nicht mit Zuverlässigkeit zu ermitteln ist. Schlußfolgerungen dürfen deshalb nur mit großer Vorsicht gezogen werden. In unserem Zusammenhang soll es nur darauf ankommen, zu zeigen, daß sowohl Isolocusbrüche als auch reziproke Chromatidentranslokationen in gleicher Weise mit zunehmender Konzentration des Agens ansteigen, so daß geschlossen werden darf, daß beiden Aberrationstypen im Prinzip gleiche Auslösungsvorgänge nach Zahl und Art zugrunde liegen dürften. Der Abfall der Kurve bei Konzentration 5×10^{-1} Mol wird durch die zunehmende Letalität bedingt sein. Andere und weitergehende Interpretationen erscheinen unsicher, solange der Wirkungsmodus des Äthylalkohols unbekannt ist.

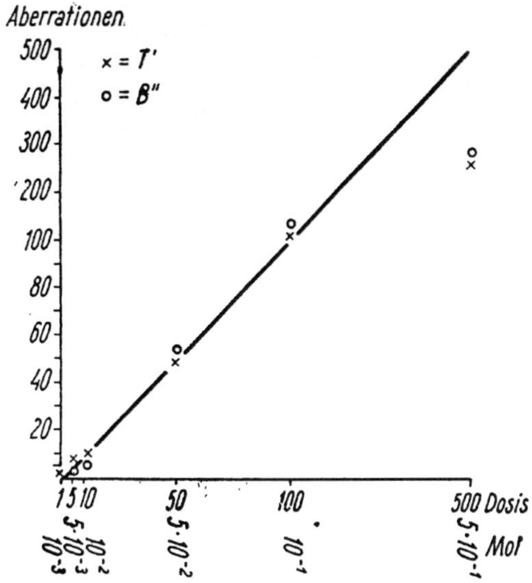

Abb. 3. Dosis-Effekt-Kurve für Isolocusbrüche und Chromatidentranslokationen nach Einwirkung der gleichen Alkoholkonzentrationen wie in Abb. 2.

Versuch 2: Der Einfluß zunehmender Behandlungszeit bei konstanter Alkoholkonzentration und variabler Erholungszeit

Der Zweck dieses Versuches war, festzustellen, wann die Chromosomenaberrationen im Zellzyklus auftreten. Zu diesem Zweck wurde bei Einwirkung einer 0,1 molaren Alkoholkonzentration in dreistündigen Intervallen während der 24stündigen Behandlungszeit und während der gleich langen Erholungszeit fixiert. Es ergibt sich aus Abb. 4a, daß der Prozentsatz aberranter Zellen während des Behandlungszeitraumes praktisch konstant bleibt und die Rate des spontanen Abweicherprozentsatzes nicht oder kaum überschreitet. Ein deutlicher Anstieg im Anteil geschädigter Zellen erfolgt erst 6–9 Stunden nach Beginn der Erholungszeit, setzt sich bis zu deren Ende fort und erreicht hier sein Maximum. Die beiden deutlichen, in allen drei Versuchswiederholungen aufgetretenen Depressionen nach 15 und 21 Stunden Erholungszeit können vorläufig noch nicht erklärt werden, sind aber möglicherweise durch Mitoseschübe bedingt. Prinzipiell das Gleiche ergibt sich, wenn die einzelnen Aberrationstypen der Auswertung zugrunde gelegt werden (Abb. 4b). Eine Ausnahme bilden hier die Duplikations-Deletionen, die von den genannten Depressionen unbeeinflußt zu bleiben scheinen. Die Anzahl der Chromatidenbrüche pro ausgewertete Zelle steigt allmählich an, während sie pro geschädigte Zelle bis zu 15 Stunden Erholungszeit annähernd konstant bleibt, um dann eine Erhöhung um ca. 0,5 Brüche pro Zelle zu erfahren (Abb. 4c).

Da erst nach etwa 9 Stunden Erholungszeit eine klare Erhöhung im Anteil geschädigter Zellen feststellbar ist, ergibt sich, daß die Effekte eindeutig während der Interphase gesetzt und keine sogenannten Primäreffekte in den in der Mitose befindlichen Zellen induziert werden. Das Auftreten der Maxima geschädigter Zellen bei 18 bzw. 24 Stunden Erholungszeit dürfte dafür sprechen, daß die Einwirkung auf die Chromosomen etwa in der ersten Hälfte der Interphase erfolgt und Zellen in einem späteren Interphasestadium unbeeinflußt bleiben. Anderenfalls wären höhere Anteile geschädigter Zellen bereits in den Fixierungen während der 24-stündigen Behandlungszeit zu erwarten.

Aus der zeitlichen Verteilung im Auftreten der einzelnen Aberrationstypen ist zu schließen, daß eine Beeinflussung der Chromosomen sowohl vor als auch während (bzw. nach) der Reduplikation nicht erfolgt, da in diesem Falle ein frühes Maximum für die sogenannten Chromatidenbrüche und ihre Reunionsprodukte und ein späteres für einen Teil der sogenannten Isolocusbrüche zu erwarten wäre, wie es für die Einwirkung von ionisierenden Strahlenarten bekannt ist. Es bleiben somit nur 2 Möglichkeiten offen: entweder werden

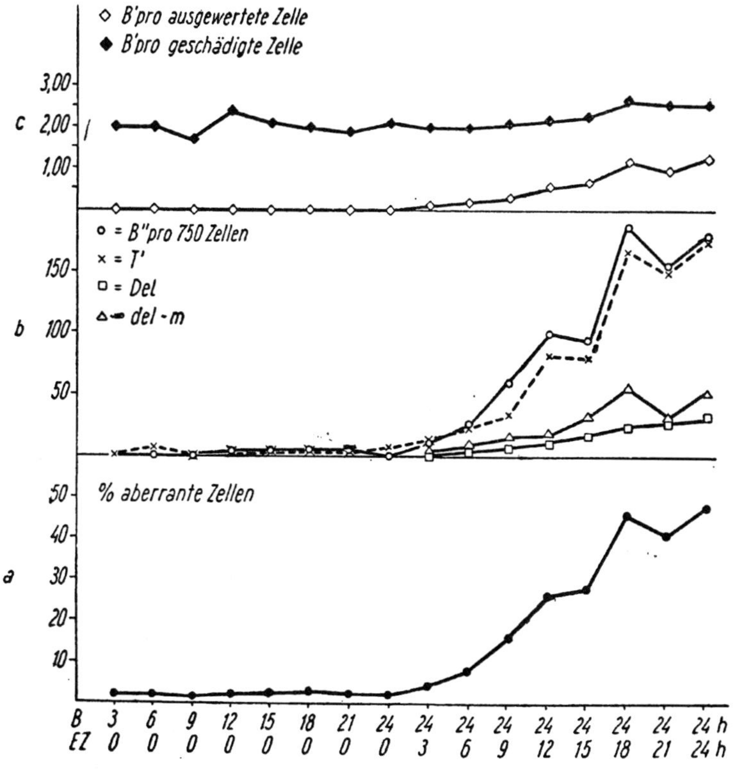

Abb. 4a—c. Der Einfluß zunehmender Behandlungszeit (*B*) bei Einwirkung von 0,1 Mol Äthylalkohol und variabler Erholungszeit (*EZ*).

a) Der prozentuale Anteil geschädigter Zellen,

b) Die absolute Anzahl von Isolocusbrüchen (*B''*), Chromatidentranslokationen (*T'*), Duplikations-Deletionen (*Del*) und Deletionen (*del-m*),

c) Die Anzahl von Chromatidenbrüchen pro ausgewertete und pro geschädigte Zelle.

die Chromosomen vor, oder sie werden während (bzw. nach) der Reduplikation (direkt oder indirekt) durch das Agens beeinflußt. Im ersten Falle müßten die „Brüche" bis zur Reduplikation offen bleiben und dürften nicht an Reunionen beteiligt werden, da alle Reunionen eindeutig auf chromatidaler Basis erfolgt waren. Es dürfte somit wahrscheinlicher sein, daß die durch den Alkohol bedingte effektive Beeinflussung der Chromosomen während oder nach der Reduplikation vollzogen wird. Die notwendige lange Behandlungszeit erweist sich bei derartigen Schlußfolgerungen naturgemäß als nachteilig und verhindert vollgültige Aussagen.

Versuch 3: Der Einfluß zunehmender Behandlungszeit bei konstanter Alkoholkonzentration und konstanter Erholungszeit

Zur Anwendung gelangte in diesem Falle ebenfalls eine 0,1 molare Alkoholkonzentration. Die Behandlungszeit erhöhte sich in 8 Intervallen von 3 bis 24 Stunden um jeweils 3 Stunden. Eine deutliche Erhöhung des spontanen Anteils geschädigter Zellen trat erst nach etwa 9-stündiger Alkoholeinwirkung ein und setzte sich dann kontinuierlich bis zur 24-stündigen Behandlungszeit fort (Abb. 5a). Das gleiche ergibt sich, wenn der Auswertung die aufgetretenen Isolocusbrüche und reziproken Translokationen zugrunde gelegt werden (Abb. 5b). Daraus ist zu entnehmen, daß die Alkoholwirkung auf die Chromosomen bei 24°C durch eine Latenzzeit ausgezeichnet ist, die sich auch durch höhere einwirkende Konzentrationen nicht

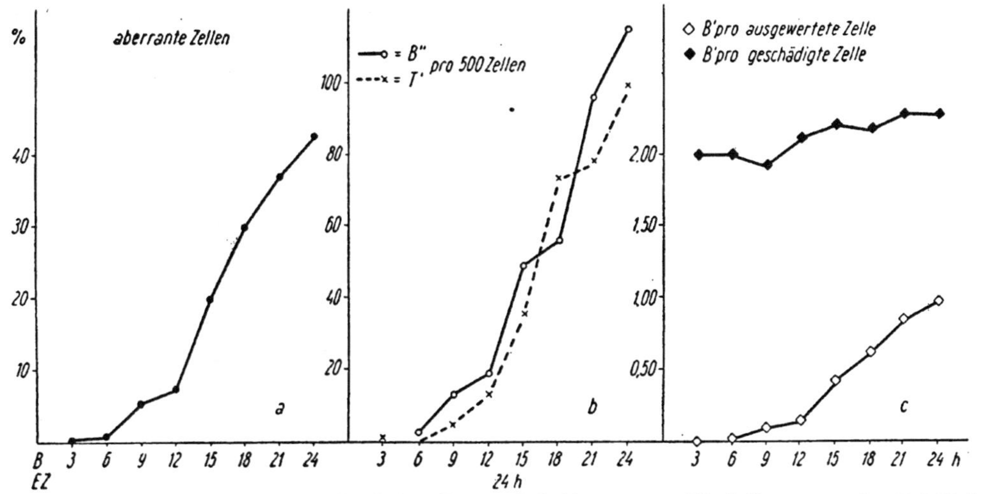

Abb. 5a—c. Der Einfluß zunehmender Behandlungszeit bei konstanter Alkoholkonzentration (0,1 Mol) und konstanter Erholungszeit.

a) Der prozentuale Anteil geschädigter Zellen,
b) Die absolute Anzahl von Isolocusbrüchen und Chromatidentranslokationen,
c) Die Anzahl von Chromatidenbrüchen pro ausgewertete und pro geschädigte Zelle.

vollständig kompensieren läßt und nicht für den Letaleffekt gilt, der nach hohen Dosen relativ schnell eintritt. Da Äthylalkohol ein sehr schnell und passiv permeierendes Agens ist, kann diese Latenzperiode mit einem zeitlich begrenzten Aufnahmevermögen des Alkohols durch die Wurzeln nicht erklärt werden, so daß sich die verzögerte Wirkung auf die effektive Beeinflussung der Chromosomen beziehen muß.

Bei kontinuierlichem Anstieg der Anzahl der Chromatidenbrüche pro ausgewertete Zelle erhöhte sich die Zahl der Chromatidenbrüche pro geschädigte Zelle, von der 9-stündigen Einwirkung angefangen, allmählich auf 2,5 Chromatidenbrüche in den beiden längsten Behandlungszeiten (Abb. 5c).

Versuch 4: Prüfung des Alkohols auf Dosisproportionalität der Wirkung bei der Induktion von Chromosomenaberrationen

Im Rahmen dieses in 2 Wiederholungen gelaufenen Versuches wurde im Konzentrationsbereich von 2 bis 0,17 Mol und entsprechenden Einwirkungszeiten an Hand des jeweiligen Anteiles geschädigter Zellen (Abb. 6a) bzw. des Anteiles von Isolocusbrüchen und Chroma-

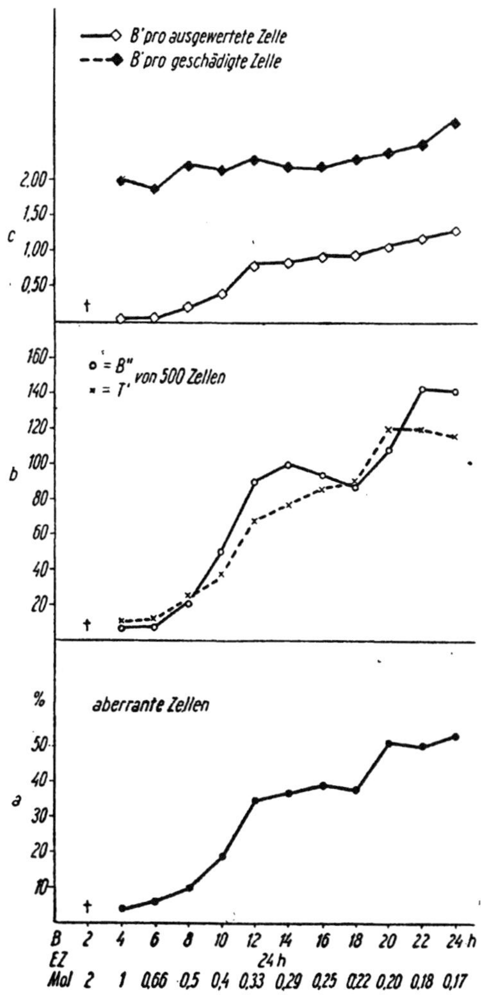

Abb. 6a—c. Prüfung des Äthylalkohols auf Dosisproportionalität der Wirkung.

a) Der prozentuale Anteil geschädigter Zellen,

b) Die absolute Anzahl von Isolocusbrüchen und Chromatidentranslokationen,

c) Die Anzahl von Chromatidenbrüchen pro ausgewertete und pro geschädigte Zelle.

tidentranslokationen (Abb. 6b) nach 24-stündiger Erholungszeit in allen Fällen auf Dosisproportionalität getestet. Die 2 molare Konzentration erwies sich bei 2-stündiger Einwirkung als letal. Bei Vorliegen einer Dosisproportionalität wäre im Bereich aller verwendeten Konzentrationen und der entsprechenden Behandlungszeiten eine identische Ausbeute an geschädigten Zellen und erfaßten Aberrationstypen zu erwarten gewesen. Statt dessen ergibt sich ein allmählicher Anstieg, der auf Grund der oben erwähnten Latenzzeit der Wirkung des Alkohols im Bereich der kurzen Behandlungszeiten gerechtfertigt wäre; im Falle der längeren Behandlung mit abnehmender Konzentration wird aber durch den quantitativen Sprung zwischen 18- und 20-stündiger Behandlungszeit bei Einwirkung von 0,22 bzw. 0,20 molarer Konzentration des Agens eine dosisproportionale Wirkung ausgeschlossen. Wenn sich eine solche auch im Bereich einiger Stufen abzeichnet, gilt eine Dosisproportionalität für den Gesamtbereich der getesteten Konzentrationen zweifellos nicht. Das wird noch deutlicher, wenn der Bewertung die Anteile an Isolocusbrüchen und Translokationen bzw. die Anzahl der Chromatidenbrüche pro geschädigte Zelle zugrunde gelegt werden (Abb. 6c). Dabei steigt besonders die Anzahl der Chromatidenbrüche pro geschädigte Zelle mit zunehmender Behandlungszeit und abnehmender Konzentration deutlich an.

Die nächsten Versuche dienten dem Zweck, durch modifizierte Milieubedingungen während der Einwirkungszeit weitere Einblicke in den Wirkungsmodus des Äthylalkohols als radiomimetischem Agens zu erlangen.

Versuch 5: Der Einfluß verschiedener Temperaturen während der Behandlungszeit mit 0.1 molaren Alkoholkonzentrationen

Während abgesehen vom 1. Versuch alle Einwirkungen des Äthylalkohols auf die Primärwurzeln bei 24°C im Thermostaten erfolgten, wurden hier Temperaturen von 6, 12, 18, 24, 30 und 36°C während der Behandlungszeit von 24 Stunden geboten. Die 24-stündige Erholungszeit erfolgte wie üblich bei 24°C. Aus der Abb. 7 ergibt sich mit Deutlichkeit ein sehr hoher Temperaturkoeffizient der radiomimetischen Wirkung des Äthylalkohols, der in dieser krassen

Form bisher für kein anderes daraufhin untersuchtes radiomimetisches Agens bekannt geworden ist. Bis zu einer Temperatur von 18° bleibt der Anteil geschädigter Zellen unter 10%, um bei 24° auf rund 40% und bei 30° auf 80% anzusteigen (Abb. 7a). Der zahlenmäßige Anteil von Isolocusbrüchen und Translokationen stieg mit der Temperaturerhöhung von 24 auf 30°C noch steiler an, da sich der Anteil an Aberrationen pro geschädigte Zelle relativ stark erhöhte (Abb. 7b). Temperaturen von 36°C im Behandlungszeitraum lagen dicht unter der Letalitätsgrenze. Der Abfall im Anteil an geschädigten Zellen ist hier mit großer Wahrscheinlichkeit darauf zurückzuführen, daß eine relativ große Anzahl von Zellen mit Aberrationen nicht mehr in die Mitose eintreten konnte. Der Anteil von Chromatidenbrüchen pro geschädigte Zelle erhöhte sich kontinuierlich mit der von 30 auf 36°C ansteigenden Temperatur

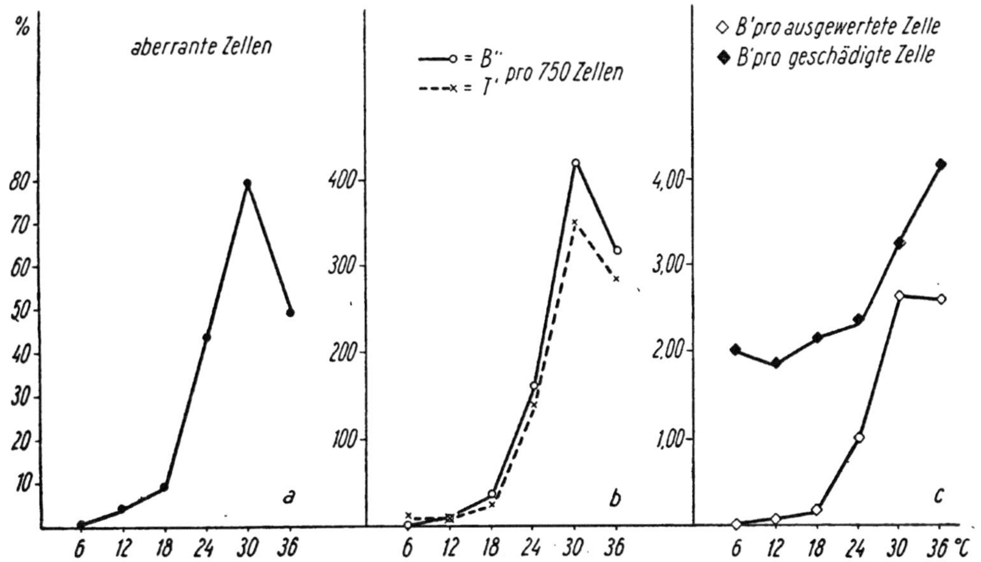

Abb. 7a—c. Der Einfluß unterschiedlicher Temperaturen während der Behandlung mit Äthylalkohol (0,1 Mol).

a) Der prozentuale Anteil geschädigter Zellen,

b) Die absolute Anzahl von Isolocusbrüchen und Chromatidentranslokationen,

c) Die Anzahl von Chromatidenbrüchen pro ausgewertete und pro geschädigte Zelle.

weiter (Abb. 7c). Außer einer erhöhten Anzahl von geschädigten Zellen wurde mit ansteigender Temperatur somit auch die Anzahl der Chromosomenaberrationen pro geschädigte Zelle erhöht. Die Zahl der Chromatidenbrüche stieg dabei auf das Doppelte, wenn die Temperatur von 24 auf 36°C erhöht wurde.

Auf Grund des Permeationsverhaltens von Äthylalkohol ist es entgegen unseren früheren Annahmen (RIEGER u. MICHAELIS 1959) nicht sehr wahrscheinlich, daß der Temperatureffekt allein auf ein mit steigender Temperatur erhöhtes Permeations- und Diffusionsvermögen des Agens zurückzuführen ist. Wenn das auch eine gewisse Rolle spielen kann, dürfte sich gleichzeitig die Beeinflussung eines temperatursensitiven chemischen Prozesses, der an der radiomimetischen Wirkung des Alkohols beteiligt ist, andeuten.

Versuch 6: Der Einfluß von 2,4 – Dinitrophenol (DNP) auf die radiomimetische Wirkung von Äthylalkohol

Behandelt wurde bei 3 Wiederholungen in 3 Parallelserien 24 Stunden lang mit 10^{-1} Mol Alkohol, 10^{-1} Mol Alkohol + 10^{-4} Mol DNP bzw. 10^{-4} Mol DNP ohne gleichzeitige Alkoholgabe. Die Erholungszeit betrug 24 Stunden. DNP allein erhöhte die spontane Aberrationsrate nicht oder nur ganz unwesentlich, während sich bei Kombination mit Äthylalkohol eine deutliche Herabsetzung im prozentualen Anteil induzierter geschädigter Zellen bzw. in der Anzahl von Isolocusbrüchen und Translokationen ergab. Die Depression machte rund 50 % des Effektes 10^{-1} molarer Alkoholkonzentrationen aus (Abb. 8).

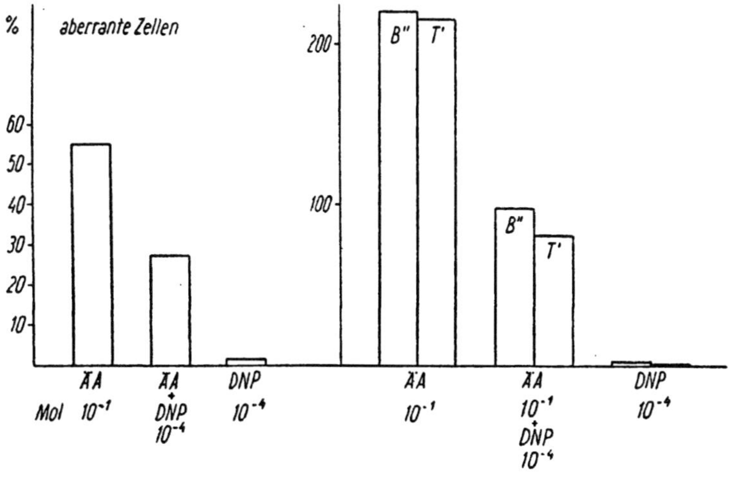

Abb. 8. Der Einfluß kombinierter Behandlung der Primärwurzeln mit 2-4-Dinitrophenol (10^{-4} Mol) und Äthylalkohol (10^{-1} Mol).

Es handelt sich hierbei vorläufig noch um einen Vorversuch, der durch Verwendung anderer DNP-Konzentrationen und durch Vor- bzw. Nachbehandlung ergänzt werden soll. Das Ergebnis des Versuches deutet an, daß im Falle der radiomimetischen Wirkung von Äthylalkohol oxydative Phosphorylierungen beteiligt sind, da DNP die Phosphorylierungen entkoppelt. In die gleiche Richtung deutet die Tatsache, daß Alkoholeinwirkung auf die Primärwurzeln in einer Stickstoffatmosphäre, d. h. bei zumindest partieller Anaerobiose im Behandlungszeitraum, die radiomimetische Wirkung des Alkohols stark herabsetzt bzw. vollständig verhindert, ohne das Aufnahmevermögen der Wurzeln für den Alkohol entscheidend zu beeinflussen. Anaerobiose wirkt in diesem Falle noch wesentlich drastischer als die 10^{-4} molare DNP-Konzentration in Kombination mit Alkohol. Im Zusammenhang mit dieser Fragestellung sollen noch mit anderen Inhibitoren Aufschlüsse über die Wirkungsweise des Agens gesucht werden.

3. *Schlußfolgerungen*

Aus den Versuchsergebnissen zeigt sich, daß Äthylalkohol bei *Vicia faba* durch eine deutliche radiomimetische Wirkung ausgezeichnet ist. Voraussetzung dafür sind bei Temperaturen von 24° C relativ lange Einwirkungszeiten und hohe Konzentrationen. Der radiomimetische Effekt stellt keine „Massenwirkung" dar und ist leicht durch den stoffwechselphysiologischen

Zustand der Zelle beeinflußbar. Der Alkohol gehört somit nach der Einteilung von Swanson und Merz (1959) in die Gruppe der nur schwach reaktiven Radiomimetica, die in fast allen Fällen durch das Auftreten eines energiebedürftigen, oxydativen Reaktionsschrittes im Wirkungssystem gekennzeichnet sind, der unbeeinflußt bleiben muß, wenn (sicherlich auf indirektem Wege) Chromosomenaberrationen ausgelöst werden sollen. Die Aberrationen werden in unserem Falle wahrscheinlich entweder ausschließlich vor oder während bzw. nach der Reduplikation im Interphasekern induziert. Die zweite Möglichkeit erscheint plausibler, und es könnte sich hier eine Beziehung zur DNS-Synthese (oder Proteinsynthese) andeuten. Die notwendigen langen Einwirkungszeiten erschweren aber die Schlußfolgerungen. Die radiomimetische Wirkung ist weiterhin durch eine weitgehend konzentrationsunabhängige Latenzphase von mehreren Stunden gekennzeichnet, vor deren Ende nach entsprechend langer Erholungszeit keine Aberrationen faßbar werden.

Es liegt nahe, an dieser Stelle über mögliche Wirkungsweisen des Äthylalkohols bei der Induktion der Chromosomenaberrationen zu spekulieren. Derartige Versuche sind naturgemäß mit allen Vorbehalten belastet, die sich im Falle Äthylalkohol noch dadurch vergrößern, daß relativ lange Einwirkungszeiten und hohe Konzentrationen erforderlich sind, um radiomimetische Wirkungen zu erzielen.

Alkohole sind zwar schwach alkylierende Verbindungen, es erscheint aber unwahrscheinlich, daß der Äthylalkohol über Alkylierungsvorgänge radiomimetische Wirkungen induziert, da die viel kräftiger alkylierenden Alkylhalogenide (z. B. Äthyljodid, Äthylbromid, Äthylchlorid) von uns getestet wurden und keine radiomimetische Wirkung zeigten. Darüber hinaus war es nicht möglich, durch Methylalkohol bzw. durch iso- und n-Propylalkohol Chromosomenaberrationen in nennenswertem Umfang zu induzieren. Wir möchten deshalb annehmen, daß der radiomimetischen Wirkung des Äthylalkohols nicht die alkylierenden Eigenschaften des Alkohols zugrunde liegen.

Äthylalkohol wirkt denaturierend[1]) auf Proteine, ohne sich mit diesen zu verbinden (nichtadditive Denaturierungswirkung) und macht sie reaktionsfähiger (zusammenfassende Darstellung bei Baker 1958). Diese Steigerung der Reaktionsfähigkeit ist wohl chemisch die wichtigste Wirkung des Äthylalkohols, die sich darin ausdrückt, daß ganz oder zum Teil in latenter Form vorliegende reaktive Konstituenten des Proteins (ionisierte Seitengruppen der Aminosäuren) in Wirkung treten und dann auf entsprechende Teste positiv reagieren. Damit kann eine erhöhte Verdaulichkeit durch Enzyme eintreten. Die Denaturierungskapazität ist stark temperatursensibel und durch einen hohen Temperaturkoeffizienten ausgezeichnet, womit sich eine Beziehung zur Temperatursensibilität der radiomimetischen Wirkung des Äthylalkohols herstellen läßt.

Es wäre also durchaus möglich, daß der Äthylalkohol seine radiomimetischen Effekte über die reversible Denaturierung bestimmter Proteine einleitet. Es könnte sich dabei um die Proteinanteile von Fermenten handeln, die etwa an der zur chromosomalen Reduplikation erforderlichen Vorläufersynthese beteiligt sind. Ihre Beeinflussung könnte dazu Veranlassung geben, daß bestimmte wichtige Bausteine fehlen, in unzureichender Menge zur Verfügung stehen oder qualitativ abgewandelt werden. Es würde sich hierbei um eine indirekte Wirkung des Agens handeln.

Nach Pischinger (1937) soll der Äthylalkohol eine langsame gelatinisierende Wirkung auf die Kernhistone ausüben können. In diesem Falle könnte sich eine direkte Wirkung auf die Chromosomen insofern andeuten, als nach Caspersson (1940) die heterochromatischen Ab-

[1]) Unter Denaturierung ist „jede nicht-proteolytische Modifikation der Struktur eines nativen Proteins zu verstehen, die zu definitiven Veränderungen in den chemischen, physikalischen oder biologischen Eigenschaften führt" (Neurath et al. 1944).

schnitte der Speicheldrüsenchromosomen histonähnliche Eiweiße aufweisen und das Heterochromatin in unserem Falle bevorzugt an den Aberrationen beteiligt wird. Allerdings sind eine ganze Reihe anderer Radiomimetica ebenfalls durch präferentielle Aberrationsauslösung in den heterochromatischen Bereichen der *Vicia faba*-Chromosomen ausgezeichnet (Ford 1949, Darlington und McLeish 1951, McLeish 1953, Revell 1953, Kihlman 1957). Es erscheint aus all diesen Gründen angebracht, bezüglich der Schlußfolgerungen größte Zurückhaltung zu üben, solange nicht ergiebigere und vor allem chemisch fundierte Daten zum Wirkungsmodus des Äthylalkohols als Radiomimeticum vorliegen.

Von Bedeutung erscheint an dieser Stelle noch der Hinweis, daß es sich beim Äthylalkohol um ein Stoffwechselprodukt handelt, das auch unter natürlichen Bedingungen in Perioden partieller Anaerobiose in recht erheblichen Mengen entstehen kann und somit möglicherweise einen der chemischen Faktoren darstellen kann, die ursächlich an der spontanen Mutabilität beteiligt sind, wenn sich gegenwärtig auch noch nicht sagen läßt, ob der Äthylalkohol in gleicher Weise bei anderen Objekten radiomimetisch wirkt.

Summary

1. Ethylalcohol in concentrations from 10^{-3} up to 5×10^{-1} Mol showed a clear radiomimetic effect, when primary roots of *Vicia faba* were treated for 24 hours. Besides isolocus-breaks and chromatid-translocations, duplication-deficiencies, deletions, triradials and a few free chromatid-breaks were found after varying recovery times. The aberrations induced were entirely of the chromatid type. Isolocus-breaks were predominant but closely followed by chromatid translocations. According to dose-effect-curves both aberration-types showed up to a concentration of 10^{-1} Mol an approximately linear increase; both are therefore probably induced by the same number of events.

2. According to various fixation times during treatment and recovery chromosome aberrations were induced by ethylalcohol in the first part of interphase, either before or during chromosome reduplication. The second possibility seemes to be more probable. Maxima of chromosome aberrations appeared after recovery times of 18 and 24 hours.

3. The radiomimetic effect of ethylalcohol was characterized by a latence period from 3 to 6 hours (temperature 24°C), during which time hardly any aberrations were induced.

4. The radiomimetic effect did not follow the "law of mass action", when concentrations from 2,0 to 0,17 Mol and corresponding treatment times were used.

5. The effect of ethylalcohol was greatly influenced by temperature. A sharp rise of aberrations was found by lifting temperatures from 18°C to 24°C and from 24 to 30°C. With temperatures of 36° the aberration rate dropped probably because cells most affected by treatment being killed before entering mitosis. It is suggested that the temperature effect is a real influence on a primary chemical reaction leading to the production of aberrations.

6. Combined treatment of roots with 10^{-1} Mol ethylalcohol and 10^{-4} Mol 2.4-Dinitrophenol for 24 hours significantly reduced the radiomimetic effect. By treatment under anoxia it was possible to suppress the radiomimetic effect entirely. It is concluded that oxidative processes are involved in the production of aberrations by ethylalcohol.

7. Ethylalcohol is according to our experiments a weakly reactive radiomimetic compound whose effects are much influenced by the metabolic state of the cell during time of treatment.

The radiomimetic action of ethylalcohol is interesting in so far as relatively great amounts of alcohol are produced by plants during periods of partial anaerobiosis. Alcohol may thus be a factor in spontaneous mutagenesis.

8. Lastly we dared a few speculations about possible modes of action of ethylalcohol in producing chromosome aberrations. Although a weak alkylating agent it seemes to be more probable that the alcohol acts by denaturing proteins.

Literatur

Baker, J. R., Principles of Biological Microtechnique. Methuen, London 1958.
Caspersson, T., Chromosoma (Berl.) **1**, 562, 1940.
Darlington, C. D., und J. McLeish, J., Nature (Lond.) **167**, 407, 1951.
Ford, C. E., Proc. 8th Int. Congr. Genetics, Hereditas (Suppl.) 570, 1949.
Kihlman, B. A., Journ. Biophys. and Biochem. Cytol. **3**, 363, 1957.
McLeish, J., Heredity **6**, (Suppl.) 125, 1953.
Michaelis, A., and Rieger, R., Chromosoma **9**, 514, 1958.
Michaelis, A., Ramshorn, K., und Rieger, R., Naturwiss. **46**, 381, 1959.
Neurath, H., Greenstein, J. P., Putnam, F. W., and Erickson, J. O., Chem. Reviews **34**, 157, 1944.
Pischinger, K., Ztschr. Zellforschg. mikr. Anat. **26**, 249, 1937.
Revell, S. H., Heredity **6** (Suppl.), 106, 1953.
Rieger, R., und Michaelis, A., Chromosoma **9**, 238, 1958.
Rieger, R., und Michaelis, A., Biol. Ztrbl. **78**, 291, 1959.

Unnatürliche Bausteine der Desoxyribonukleinsäure

Adolf Wacker

Organisch-chemisches Institut der Technischen Universität Berlin, Berlin-Charlottenburg

Bei Arbeiten über die Funktion der Folsäure im Bakterienstoffwechsel untersuchten wir auch bei *Sc. faecalis* R den Mechanismus der kompetitiven Enthemmung von 5-Bromuracil (BU) mit Thymin und den der nichtkompetitiven mit Folsäure. Mit Hilfe eines ^{82}Br-markierten BU fanden wir dabei 1951 (Weygand, Wacker und Grisebach 1951; Weygand, Wacker und Dellweg 1952; Weygand 1951), daß im Falle der nichtkompetitiven Enthemmung von BU mit Folsäure, BU an Stelle von Thymin in der Desoxyribonukleinsäure (DNS) verbleibt. Wie wir ebenfalls zeigen konnten, wird BU ohne das Bakterienwachstum zu hemmen in die DNS auch anderer Bakterien, wie *E. coli* oder *Enterococcus* eingebaut (Weygand und Wacker 1952). – Inzwischen haben wir uns weiter mit der Aufnahme des BU durch Bakterien beschäftigt. Mit dem Einbau anderer Thymin-analoger Verbindungen suchten wir insbesondere die Spezifität des Ersatzes von Thymin in der DNS zu klären. Darüber hinaus haben wir den Einfluß von BU auf Sulfathiazol-gehemmte Zellen von *Proteus mirabilis* untersucht und dabei auch die mutagene Wirkung des BU geprüft.

5-Bromuracil; 5-Bromcytosin[1])

Befindet sich im Nährmedium von *Enterococcus* Stei oder *E. coli* B 10 γ Thymin-[2-^{14}C] pro ml, so nehmen 100 mg Bakterien von *Enterococcus* Stei 200 γ Thymin auf und von *E. coli* B 15 γ (quantitative Bestimmung der Radioaktivität mittels Gaszählrohr). Spaltet man beide Bakterienstämme mit Perchlorsäure und bestimmt nach papierchromatographischer Trennung die Radioaktivität der Pyrimidinbasen, so ist nur Thymin radioaktiv, Uracil und Cytosin sind inaktiv. Im Vergleich zu Thymin nimmt *Enterococcus* Stei unter den gleichen Bedingungen 1250 γ Uracil-[2-^{14}C]/100 mg Bakterien auf und *E. coli* B 750 γ. Wie die Perchlorsäurespaltung der mit Uracil-[2-^{14}C] gewachsenen Bakterien zeigt, wird Uracil teilweise in Cytosin und Thymin umgewandelt. Im Gegensatz zu *E. coli* B nimmt *Enterococcus* Stei nur wenig Cytosin auf. Nach Wachstum mit 10 γ Cytosin-[2-^{14}C] hat *Enterococcus* Stei 60 γ/100 mg Bakterien aufgenommen, *E. coli* B 1250 γ. Beide Stämme wandeln das aufgenommene Cytosin teilweise in Uracil und Thymin um.

Aus früheren Untersuchungen mit 5-Bromuracil-[^{82}Br] war bekannt, daß *Enterococcus* Stei und *E. coli* B diesen unnatürlichen Baustein an Stelle von Thymin in ihre DNS einbauen (Weygand und Wacker 1952). Nach 36 Stunden Wachstum mit 10 γ 5-Bromuracil-[2-^{14}C]

[1]) Veröffentlichung in Vorbereitung.

hat *Enterococcus* Stei 750 γ/100 mg Bakterien aufgenommen und *E. coli* B 85 γ. Spaltet man beide mit 5-Bromuracil-[2-^{14}C] gewachsenen Bakterienstämme mit Perchlorsäure und isoliert papierchromatographisch die Pyrimidinbasen, so stellt man fest, daß außer 5-Bromuracil auch Uracil, Cytosin und Thymin radioaktiv geworden sind. Ungefähr 30% des aufgenommenen 5-Bromuracils wurden von *Enterococcus* Stei dehalogeniert, bei *E. coli* B sind es etwa 70%.

Kürzlich berichteten KORNBERG und Mitarb. über die enzymatische Synthese einer DNS, die an Stelle von Cytosin 5-Bromcytosin enthielt (BESSMAN, LEHMAN, ADLER, ZIMMERMANN, SIMMS und KORNBERG 1958). Bei unseren Untersuchungen fanden wir nun, daß im Gegensatz zu 5-Bromuracil-[2-^{14}C] sowohl *Enterococcus* Stei als auch *E. coli* B keine nachweisbaren Mengen 5-Bromcytosin aufnehmen. Nach Wachstum beider Stämme mit 10 γ 5-Bromcytosin-[^{82}Br] oder 5-Bromcytosin-[2-^{14}C] (spezifische Aktivität beider Verbindungen 1 mC/mMol) zeigten die Bakterien im Methandurchflußzähler eine Radioaktivität, die nur wenige Impulse über dem Nullwert lag.

Die Versuche mit 5-Bromcytosin-[^{82}Br] und 5-Bromcytosin-[2-^{14}C] zeigen deutlich, daß 5-Bromcytosin weder Cytosin noch Thymin bei *Enterococcus* Stei und *E. coli* B ersetzen kann. Was den Ersatz des Thymins betrifft, steht dies in Übereinstimmung mit dem Desoxyribonucleinsäure-Modell von WATSON und CRICK, im Falle des Cytosins aber anscheinend im Widerspruch zu der enzymatisch gebildeten 5-Bromcytosin-enthaltenden DNS (BESSMAN, LEHMAN, ADLER, ZIMMERMAN, SIMMS und KORNBERG 1958). Berücksichtigt man aber, daß in der Zelle die DNS als Nucleoprotein vorliegt, und somit das umhüllende Protein strukturell in naher Beziehung zur DNS steht, so findet man vielleicht eine Erklärung dafür, daß Cytosin als Bestandteil der Bakterien-DNS in 5-Stellung keinen Substituenten tragen kann. Die Tatsache, daß bestimmte Phagen an Stelle von Cytosin 5-Hydroxymethylcytosin (COHEN 1953) enthalten, mag mit der strukturellen und genetischen Sonderstellung der Phagen zusammenhängen.

Während Uracil und Cytosin gegenseitig ineinander umgewandelt werden, ist die Bildung des Thymins ein irreversibler Vorgang. Um so merkwürdiger ist es, daß 5-Bromuracil, das Thymin bei *Enterococcus* Stei und auch bei anderen Bakterien zu mehr als 50% ersetzen kann, in Uracil, Cytosin und damit letztlich auch in Thymin umgewandelt wird. Die Abspaltung des Bromatoms in 5-Stellung des Uracil-Moleküls ist aber nicht allein auf Bakterien beschränkt. Wie auch Untersuchungen bei Säugetieren und am Menschen (BARRETT und WEST 1958; PAHL, GORDON und ELLISON 1959) gezeigt haben, können verschieden substituierte Pyrimidine in vivo dehalogeniert werden.

Wuchsstoffaktivität und mutagene Wirkung von 5-Bromuracil[1])

Im Zusammenhang mit der Tatsache, daß Thymin die Sulfathiazol-Hemmung bei *Proteus* nur halbmaximal aufzuheben vermag (MÖLLER, WEYGAND und WACKER 1950), untersuchten wir den Einfluß von BU auf Sulfathiazol-gehemmte Zellen von *Proteus mirabilis*. Dabei prüften wir auch die mutagene Wirkung des BU. Die Mutagenität von BU wurde bei *Proteus mirabilis* VI/Sb-3 bei dem Mutationsschritt Streptomycin-Abhängigkeit → Streptomycin-Unabhängigkeit getestet und mit der Wirkung eines bekannten mutagenen Agens, $MnCl_2$, verglichen. Neben nichtradioaktivem BU wurde auch ^{14}C-markiertes und ^{82}Br-markiertes BU verwendet.

---

1) A. WACKER und H. BÖHME, Z. Naturforschung 14b, 703—706 (1959).

Der Streptomycin-abhängige Stamm *Proteus mirabilis* VI/Sb-3[1]) wird durch 15 γ Sulfathiazol pro ml Agar mäßig in seinem Wachstum gehemmt. 10 γ Thymin heben teilweise diese Hemmung auf. Interessanterweise hat BU die gleiche Wirkung wie Thymin. Auch in flüssigem Medium hebt BU halbmaximal die Sulfathiazol-Hemmung auf. Die gleiche Wirkung wie BU hat auch 5-Chloruracil. Die enthemmende Wirkung des BU läßt sich auch bei dem Stamm *Proteus mirabilis* VI nachweisen, der kein Streptomycin als Wuchsstoff benötigt. – Wie frühere Untersuchungen über den Wirkungsmechanismus des Sulfathiazols mit ^{14}C-markierten Verbindungen gezeigt haben, blockiert Sulfathiazol die Synthese der Folsäure und damit verbunden die der Purine und des Thymins (WACKER, TREBST und SIMON 1957; WEYGAND, WACKER, TREBST und SWOBODA 1954). Es ist daher nicht verwunderlich, daß Thymin in Anwesenheit von Purinen die Wachstumshemmung aufhebt. Die Tatsache, daß BU die gleiche Wirkung wie Thymin hat, läßt sich nur so erklären, daß diese Verbindung die Funktion des Thymins in der Zelle übernimmt. Den Nachweis, daß BU tatsächlich von den Zellen aufgenommen wird, erbrachten wir durch die Verwendung von radioaktiv-markiertem BU.

Nachdem bewiesen war, daß BU an Stelle von Thymin in die DNS von *Proteus mirabilis* eingebaut wird, wurde geprüft, ob dieser Einbau eine Erhöhung der Mutabilität zur Folge hat. Die Zahl der spontan auftretenden Streptomycin-unabhängigen Mutanten beträgt unter den in diesen Versuchen angewandten Bedingungen bei dem Stamm VI/Sb-3 20–40/10^8 plattierter Zellen. Unter dem Einfluß von $MnCl_2$ steigt die Mutantenhäufigkeit auf 500 bis 1000/10^8 an. Im Gegensatz dazu konnte nach Zugabe von BU und gleichzeitiger Thymin-Hemmung durch Sulfathiazol eine signifikante Steigerung der Mutantenhäufigkeit nicht beobachtet werden.

Wie frühere Versuche zeigten, nehmen Bakterien auch in ruhendem Zustand BU auf (WEYGAND, WACKER und GRISEBACH 1951; WEYGAND, WACKER und DELLWEG 1952; WEYGAND 1951). Das gleiche trifft für *Proteus mirabilis* zu. Nimmt *P. mirabilis* in Phosphat-Puffer suspendiert radioaktives 5-Bromuracil-[^{82}Br] auf, so führt dies ebenfalls nicht zu einer Erhöhung der Mutantenhäufigkeit. Dieser Befund ist von besonderem Interesse, da nämlich ^{82}Br nach 36 Stunden in das Edelgas Krypton zerfällt und außerdem eine harte β- und γ-Strahlung aussendet. Das in die DNS eingebaute 5-Bromuracil-[^{82}Br] muß bei dem Zerfall des ^{82}Bromatoms an dieser Stelle der DNS durch Entstehung von Uracil oder eines anderen Moleküls eine falsche Information liefern.

Die mutagene Wirkung von BU wurde auch von anderer Seite bei Bakterien untersucht. In Übereinstimmung mit den Ergebnissen von NOVICK und SZILARD (NOVICK und SZILARD 1951) sowie ZAMENHOF et al. (ZAMENHOF, DE GIOVANNI und RICH 1956) bei *E. coli* fanden wir bei *P. mirabilis* ebenfalls keine Erhöhung der Mutationsrate, gleichgültig ob inaktives oder radioaktives BU verwendet wurde. Wie die vorliegenden Versuche gezeigt haben, kann BU als Wuchsstoff Thymin bei Sulfathiazol-gehemmten Zellen ersetzen und auch strukturell den Platz des Thymins in der DNS der Bakterienzelle einnehmen. Da nach unserer heutigen Auffassung die DNS der potentielle Träger der genetischen Information ist, hätte man zumindest bei dem Versuch mit 5-Bromuracil-[^{82}Br] erwarten können, daß sich die durch den Einbau ergebenden Veränderungen genetisch auswirken. Das Ausbleiben dieser Wirkung an dem von uns untersuchten Mutationsschritt schließt jedoch die Möglichkeit einer mutativen Veränderung an einer anderen Stelle des genetischen Materials nicht aus.

Der Befund, daß Thymin als Wuchsstoff bei Bakterien durch BU ersetzt werden kann, hat eine Parallele in dem Ersatz von Thymindesoxyribosid durch 5-Bromuracildesoxyribosid bei den Desoxyriboside-bedürftigen Bakterien *Lb. leichmannii* und *Lb. acidophilus* R_{26} (WACKER, TREBST und WEYGAND 1956).

---

[1]) Dieser Stamm wurde im Laufe genetischer Untersuchungen von H. BÖHME isoliert.

5-Nitrouracil; 5-Aminouracil; 2-Thiothymin[1])

5-Bromuracil hebt bei Bakterien nicht nur die 5-Nitrouracil-Hemmung auf (WEYGAND und WACKER 1952), sondern auch die des 5-Aminouracils.[2]) Mit ^{14}C-markiertem 5-Nitrouracil und 5-Aminouracil wurde die Aufnahme und der Stoffwechsel dieser Verbindungen sowie der Mechanismus der kompetitiven Enthemmung näher untersucht. 2-Thiothymin beeinflußt in gleicher Weise wie 5-Aminouracil den Stoffwechsel der DNS bei bestimmten Bakterien (DUNN und SMITH 1955). In Anwesenheit dieser beiden Pyrimidin-analogen Verbindungen findet man in der DNS zusätzlich 6-Methyladenin. Im Rahmen der Untersuchungen über 5-Aminouracil-[2-^{14}C] haben wir uns ebenfalls mit der Aufnahme und dem Stoffwechsel von 2-Thiothymin-[2-^{14}C] beschäftigt.

Wie aus Tab. 1 hervorgeht, nehmen alle untersuchten Stämme 5-Nitrouracil, 5-Aminouracil und 2-Thiothymin aus dem Nährmedium auf. Auffällig ist, daß die beiden *E. coli*-Stämme im Vergleich zu *Enterococcus* Stei und *Streptococcus faecalis* R wesentlich weniger 5-Aminouracil und 2-Thiothymin aufnehmen. Bei 5-Nitrouracil ist die unterschiedliche Aufnahme nicht so ausgeprägt.

Tabelle 1. Aufnahme verschiedener Pyrimidine durch *Enterococcus* Stei, *Streptococcus faecalis* R., *E. coli* B und *E. coli* 1883 Co nach 36 Std. Wachstum

Im Nährmedium vorhanden 10 γ/ml	Aufnahme der radioaktiven Verbindung γ/100 mg Bakterien*)			
	Enterococcus Stei	*Streptococcus faecalis* *R*	*E. coli* B	*E. coli* 1883 Co
5-Nitrouracil-[2-^{14}C]	15	12	8	7
5-Aminouracil-[2-^{14}C]	35	40	5	6
2-Thiothymin-[2-^{14}C]	40	25	2	2

*) Quantitative Messung von ^{14}C in der Gasphase[9].

Wächst *Enterococcus* Stei entweder mit 5-Nitrouracil-[2-^{14}C], 5-Aminouracil-[2-^{14}C] oder 2-Thiothymin-[2-^{14}C], so verringert sich die Radioaktivität der Bakterienzellen, wenn gleichzeitig Thymin oder 5-Bromuracil vorhanden ist. 10 γ Thymin erniedrigen die Aufnahme von 10 γ 5-Nitrouracil um mehr als 80%, 10 γ 5-Broumuracil um etwa 70%. Bei 5-Aminouracil-[2-^{14}C] und 2-Thiothymin-[2-^{14}C] ist Thymin im Vergleich zu 5-Bromuracil wirksamer bei der Verdrängung der unnatürlichen Pyrimidine. Umgekehrt wird die Aufnahme radioaktiven Thymins nur durch 5-Bromuracil verringert, nicht aber durch 5-Nitrouracil, 5-Aminouracil oder 2-Thiothymin.

Untersucht man in den mit 5-Nitrouracil-[2-^{14}C] gewachsenen Bakterien die Verteilung der Radioaktivität, so zeigt sich, daß mehr als 90% der Radioaktivität nicht in der DNS enthalten ist. Nach Spaltung der DNS mit Perchlorsäure findet man auf dem Papierchromatogramm deutlich nur eine radioaktive Substanz. Infolge des äußerst geringen Vorkommens der radioaktiven Verbindung in der DNS konnte vorerst nur durch einen Vergleich der R_f-Werte des 5-Nitrouracils und seiner biochemisch möglichen Umwandlungsprodukte auf die Struktur

[1]) Veröffentlichung in Vorbereitung.

[2]) Unveröffentlichte Ergebnisse.

der radioaktiven Verbindung geschlossen werden. Danach ist die aus der DNS isolierte radioaktive Substanz mit 5-Nitrouracil nicht identisch.

Bei den mit 5-Aminouracil-[2-^{14}C] gewachsenen Bakterien befindet sich der größte Teil der Radioaktivität ebenfalls nicht in der Bakterien-DNS. Nach der Perchlorsäurespaltung der DNS sind auf den Papierchromatogrammen in verschiedenen Lösungsmittelsystemen vier Flecke radioaktiv, wovon drei in ihren R_f-Werten mit 5-Aminouracil, Cytosin und Thymin übereinstimmen. Die vierte radioaktive Substanz haben wir in ihrer Struktur noch nicht aufgeklärt. Spaltet man die gesamten Bakterien mit Perchlorsäure, so tritt zusätzlich noch Uracil und eine weitere unbekannte Verbindung auf, die den größten Teil der Radioaktivität ausmacht.

Enterococcus Stei nimmt etwa die gleiche Menge 2-Thiothymin auf wie 5-Aminouracil. Da bei der Perchlorsäurespaltung 2-Thiothymin in Thymin umgewandelt wird, war es nicht möglich, auf diese Weise die Struktur der radioaktiven Verbindung in den Bakterien aufzuklären. Es wurden daher aus den mit radioaktivem 2-Thiothymin gewachsenen Zellen einmal die Desoxyriboside aus der DNS nach enzymatischer Spaltung isoliert, zum anderen die mit NaCl isolierte DNS mit Ameisensäure gespalten. In beiden Fällen findet man nur eine radioaktive Substanz, die mit Thymin-desoxyribosid bzw. Thymin identisch ist. Nach Perchlorsäurespaltung der gesamten Bakterien findet man ebenfalls keine andere radioaktive Base außer Thymin.

Wie die Versuche zeigten, werden 5-Nitrouracil, 5-Aminouracil und 2-Thiothymin von den Bakterien nur sehr wenig aufgenommen. Im Gegensatz zu 5-Bromuracil, das Thymin in der DNS weitgehend ersetzen kann, wird 5-Aminouracil nur zu etwa 0,1 % an Stelle von Thymin in die DNS eingebaut. 5-Nitrouracil und 2-Thiothymin werden als solche nicht in die DNS aufgenommen. Das äußerst geringe Vorkommen von 5-Aminouracil in der Bakterien-DNS und die vollständige Umwandlung von 5-Nitrouracil und 2-Thiothymin in andere Pyrimidinbasen erklären auch die bisher in der Literatur mitgeteilten negativen Befunde über die Aufnahme und den Einbau dieser unnatürlichen Pyrimidine (BENNETT 1952).

Literatur

BARRETT, H. W., und WEST, R. A., J. Am. Chem. Soc. 78, 1612 (1958).
BENNETT, L. L., jr., J. Am. Chem. Soc. 74, 2432 (1952).
BESSMAN, M., LEHMAN, J. R., ADLER, J., ZIMMERMAN, S. B., SIMMS, E. S., und KORNBERG, A., Proc. Natl. Acad. Sci. U. S. 44, 633 (1958).
COHEN, S. S., Cold Spring Harbor Symp. Qant. Biol. 18, 221 (1953).
DUNN, B. D., und SMITH, J. D., Nature, 175, 336 (1955).
MÖLLER, E. F., WEYGAND, F., und WACKER, A., Z. Naturforschg. 5b, 18 (1950).
NOVIK, A., und SZILARD, L., Cold Spring Harbor Symp. 16, 337 (1951).
PAHL, H. P., GORDON, M. P., und ELLISON, R. R., Arch. Biochem. Biophys. 79, 245 (1959).
WACKER, A., TREBST, A., und WEYGAND, F., Z. Naturforschg. 11b, 7 (1956).
WACKER, A., TREBST, A., und SIMON, H., Z. Naturforschg. 12b, 315 (1957).
WEYGAND, F., WACKER, A., und GRISEBACH, H., Z. Naturforschg. 6b, 177 (1951).
WEYGAND, F., in Proceedings of the Isotope Techniques Conference, Oxford, July 1951.
WEYGAND, F., WACKER, A., und DELLWEG, H., Z. Naturforschg. 7b, 19 (1952).
WEYGAND, F., und WACKER, A., Z. Naturforschg. 7 b, 26 (1952).
WEYGAND, F., WACKER, A., TREBST, A., und SWOBODA, O. P., Z. Naturforschg. 9b, 764 (1954).
ZAMENHOF, S., DE GIOVANI, und RICH, R., J. Bacteriol. 71, 60 (1956).

Some observations on the interaction between alkylating agents and phage and their relevance to problems of chemical mutagenesis

A. Loveless

Chester Beatty Research Institute (Institute of Cancer Research: Royal Cancer Hospital), London, S. W. 3

1. Some cautionary remarks

For many reasons it is impossible to make quantitative comparisons between different alkylating agents in respect of their biological, and especially mutagenic, activity. Consequently, the use of expressions such as "good mutagen" may be misleading insofar as they may refer to chemical desiderata unrelated to the potential mutagenic efficiency of a particular compound. Apart from such obvious differences as reactivity, solubility, volatility, etc., which may clearly affect the quantitative efficiency of a compound, the undermentioned properties must be borne in mind in any experiment involving the types of compounds named:

1. Sulphur mustards: are very rapidly hydrolysed, yielding acid, in aqueous media, e. g., mustard gas (di (2-chloroethyl) sulphide) is 50% hydrolysed at pH 7 and 37° C. in ca. 5 min.

2. Nitrogen mustards:

   a) Are unreactive if the chloroethyl-substituted N atom is ionized: the proportion in the cationic form at a given pH will depend upon the pK of the amine. Compounds of this type may be supplied as hydrochlorides, in which case it may be necessary to neutralize the solution.

   b) Aliphatic types rapidly cyclize to the quaternary ethylenimmonium ion (e. g., HN2 (di(2-chloroethyl)methylamine) 50% in 2 min. at 37° C. and pH 7). This form is chemically more reactive but may not readily penetrate certain membranes as we have shown (1, 2) for certain bacteria and phages.

   c) If employed for long periods, or if solutions are used after prolonged standing, hydrolysis may diminish the effective concentration and acid liberated may reduce the pH. NH2 suffers 50% hydrolysis in 131 min. at pH 7 and 37° C. (W. Davis, personal communication.) Strong solutions of hydrochlorides are relatively stable owing to low pH, but cyclization and subsequent hydrolysis will nevertheless proceed slowly.

3. Ethylenimines:

a) May be strongly basic and so give alkaline solutions which are less reactive. This does not apply to amide types, i. e., N-acyl ethylenimines.

b) Basic ethylenimines will be partially in the cationic form at neutrality, depending upon pK. Thus, ethylenimine itself, pK 7.88, is approximately 89%, ionized at pH 7. Whilst more reactive in this form, they may, like the quaternary transformation products of nitrogen mustards, find greater difficulty in penetration.

c) Solutions are liable to deteriorate either through hydrolysis or polymerisation. These, like other reactions of ethylenimines, are acid-catalyzed.

4. Alkyl sulphates: are rapidly hydrolyzed in aqueous media yielding strongly acid alkyl hydrogen sulphates.

5. Esters of alkyl sulphonic acids: hydrolyze, in general, much more slowly than alkyl sulphates, e. g., unsubstituted straight chain alkyl esters of methane sulphonic acid have a half-life for hydrolysis at pH 7 and 37° C. of about 8 h. Care should be observed, however, with g-substituted esters such as dimethyl myleran, half-life ca. 15 min. [3].

6. Epoxides: are the least difficult to handle; hydrolysis is slow, e. g., diepoxybutane has a half-life of ca. 100 h. at 37° C., and no change in pH is involved. Strong solutions of the liquid types may, however, have untoward solvent effects [1]. Propiolactone may be considered as an epoxide for practical purposes.

These cautionary remarks are not based solely on chemical observations, but have been borne out in experiments with bacteria and with various phages [1, 2, 4]. Whilst they serve to illustrate the very great difficulty – if not impossibility – of making quantitative comparisons between alkylating agents as mutagens, they do not give any basis for qualitative differences, i. e., specificity. Results obtained in a study of the inactivation of phage by alkylating agents have, however, revealed certain facts which might relate thereto.

2. *Inactivation of phage by alkylating agents*

The main body of our work in this field is either published or in press [1, 2, 4, 5, 6]; consequently only certain aspects will be dealt with here. Firstly, it has been established that the inactivation of phage, as an infective particle, by alkylating agents derives from reaction with the 'germinal substance' [7], and it seems inescapable that this means reaction with the DNA. Secondly, it has been established that the consequences of this reaction are not fully realised until sometime after the primary chemical reaction (alkylation) has been effected. In the case of compounds subtending more than one alkylating group, inactivation appears to be coincident with the reaction of a second alkylating centre in the same molecule. This may occur some time later than reaction of the first group and a period of post-treatment decay is therefore observed. Also, if phage synthesis is allowed to supervene, by rapid adsorption of the treated phage to sensitive bacteria in nutrient medium, the infectivity of a large proportion of the treated particles may be preserved. This observation may have importance where polyfunctional compounds have been employed as mutagens under conditions where replication may have preceded completion of the reaction; i. e., it might be that polyfunctional compounds have been used effectively only as monofunctional compounds. The effect may also be responsible for the observation [8] that the earlier part of the interkinetic period is most sensitive to radiomimetic agents. Our results with phage indicate that, if the reaction is allowed to go to completion, exponential survival curves are obtained. In the case of monofunctional compounds the situation is more complex. Monofunctional alkylation of phage germinal substance does not render the phage non-infective at the time of reaction, but produces an unstable adduct which subsequently undergoes spontaneous reaction – decomposition or re-arrangement – observed as a gradual loss of infectivity, which is dose and pH-dependent. There is thus also a 'decay' period following reaction with monofunctional agents which is more prolonged and due to different reasons than that obtaining with polyfunctional compounds. The dose-dependence of the decay rate was earlier [4] held to be the reason for the 'multi-hit' type of curve obtained for phage inactivation by monofunctional compounds. It must be added, however, that the inactivation of bacteria by monofunctional alkylating agents is determined more by dose-rate – i. e., the absolute concentration of the alkylating agent during treatment – than by dose – i. e., the product of the concentration x the time of treatment. No explanation can presently be offered for this fact, but it is evidently an additional factor affecting the quantitative comparison of the biological efficiency of alkylating agents.

The "decay pattern" is not the same for all compounds, whether mono- or polyfunctional, and might be considered as reflecting a different degree of chemical stability of various alkylation adducts. The implicit suggestion is that specificity of mutational activity derives not from specificity on the part of the mutagen, nor on the part of the sensitive site, but from the properties of the primary reaction product. Especially might this be significant if the scoring procedure permits some selection between rapid and delayed effects.

3. Pre-mutation and the role of 'loops'

It is obvious that in the chemical mutation of phage, be it by bromouracil incorporation [9], ethylation [10], or deamination by nitrous acid [11] or other means, the primary chemical reaction cannot constitute the mutation (unless it derives from deletion), since the latter is observed as a mutant plaque containing millions of phages synthesized in the absence of the mutagen. Since, in the three examples cited, it has been established that the primary action of the mutagen is to modify the phage DNA, it must be presumed that the mutation arises subsequently as an 'error of replication'. In our own experiments with ethylation, it was observed by the single burst method that bacteria giving mutants also yielded wild-type phage and that the relative proportions varied widely. There was thus no reason to attribute the production of both types to chemical modification of only one strand of a parental double helix. There is no ground for assuming at present that in organisms the primary site of reaction of alkylating agents is the DNA, although a priori this seems very likely, nor that such a reaction would predispose the DNA to mutational change by 'errors of replication'. We hope to investigate this possibility in bacteria by recombination or transduction methods. If the primary reaction is of this kind in organisms, then the remarks made above in reference to phage must equally apply.

Certain other observations on the inactivation of phage by alkylating agents indicate that such 'errors of replication' might arise in a fashion analogous to that proposed by REVELL (cf. this Symposium) for chromosomes but at a molecular level. These observations showed that the modification of phage DNA following alkylation, referred to above as decay, proceeds only when the DNA is tightly coiled as in the phage head. This was demonstrated in the following ways:

1. Phage treated briefly at high concentration of reagent and adsorbed rapidly to washed cells gave infective centres which underwent very little decay on subsequent storage in carbon-free or cyanide-containing media. For technical reasons the only monofunctional compound studied was ethylenimine.
2. Similarly treated phage subjected to immediate osmotic shock gave a DNA solution of undiminished relative viscosity which remained stable during 24 h. at room temperature. When stored intact for 24 h. prior to shock the phage gave a solution of greatly diminished viscosity. The measurements were made after proteolytic digestion and the completeness of the shock was confirmed by electron microscopy. Again for technical reasons, the only monofunctional compound studied was ethylene oxide.

The greater sensitivity of a coiled structure toward bifunctional reagents could be readily comprehended on the basis of cross-linkage, but this is exceedingly unlikely in the case of ethylene oxide or ethylenimine and quite impossible for compounds inserting unsubstituted alkyl groups. It must be supposed that alkylated DNA is intrinsically less stable in a folded form and from this it must follow that regions of coiling or loops will be less stable than extended or non-overlapping regions. It is difficult not to recall the analogous situation in the resting stage heterochromatic regions of chromosomes discussed by OCKEY (this Symposium). Now, of course, these observations refer only to inactivation of phage, but the evidence that certain regions may be more sensitive than others owing to geometrical configuration alone may have some relevance to the problem of specificity, which, so far as alkylating agents are concerned, seems otherwise incomprehensible. Even more hypothetically, we might suppose that a mechanism such as that proposed by REVELL, operating at the molecular level, might produce mutation during replication by intramolecular translocation. Such a change would conserve all the materials for back-mutation and provide a basis for pleiotropy.

4. *Ethyl methane sulphonate – a 'good mutagen'*

Amongst alkylating agents, ethyl methane sulphonate does seem to warrant the term 'good mutagen'. It is probably in essence no better than diethyl sulphate, but has the practical advantages of greater solubility and slower hydrolysis rate. Although the latter means that other types of reaction will be correspondingly slower than those of diethyl sulphate, it also means that untoward side effects due to acid are more likely to be avoided. Either or both of these substances have been employed as mutagens in *Drosophila* [12, 13], *Neurospora* [14 and cf. WESTERGAARD, this Symposium], phage T 2 [10, 6], T 4 (D. M. GREEN, personal communication) and barley [15, 16]. The dramatic results obtained with *Neurospora*, phage and barley prompted Dr. Sheila HOWARTH and myself to investigate the mutagenic action of ethyl methane sulphonate in bacteria. A report of these experiments is now in press but a summary of the findings may be of interest. The systems studied were:

1. Revertant and supressor mutations in *Salmonella typhimurium* LT 2, cystine requiring;
2. "Forward" mutation in the same organism and
3. Mutation to *T 6* resistance in *E. coli B* and *B/r*.

At approximately 70% survival the yield of revertants was ca. 100 x and of suppressors ca. 1000 x spontaneous. Five forward mutants were isolated from about 1000 colonies replica-plated, each having a different nutritional requirement. For *T6* resistance, a yield approaching 100 x spontaneous was obtained at 70% survival with *B/r*, whilst a comparable value was only achieved with *B* at 20% survival. It may be noted in passing that *B/r* is more resistant to a variety of mono- and bifunctional alkylating agents than *B* in respect of colony-forming ability. The comparable mutability of the two strains argues against any theory that *B/r* possesses a biochemical reparative mechanism absent in *B*, and suggests that this may also be the case where radiation is concerned. In this connection we have found that the induction of mutation of *T2* by ethyl methane sulphonate proceeds equally well whether the treated virus is grown in *B* or *B/r*.

The results obtained with alkylating agents as chemical mutagens and with ethyl methane sulphonate in particular in a wide variety of systems indicate that a common mode of action should be sought in viruses, bacteria and higher organisms, contrary perhaps to the view expressed by Professor WESTERGAARD in this Symposium. With regard to the possible role of 'decay' processes in relation to specificity, it must be noted that no difference was observed in the proportion of mutants derived from ethylated *T2* when plated before and after decay. However, only the *r* phenotype has been scored and no attempt made to map the mutations obtained.

Finally, given a good mutagen, it should be tested for ability to evoke any biological consequence for which a mutational origin has been postulated – e. g., carcinogenesis. We have so far failed to induce *E. coli* strain *K12* (λ) with ethyl methane sulphonate.

This work has been supported by grants to the Chester Beatty Research Institute (Institute of Cancer Research: Royal Cancer Hospital) from the British Empire Cancer Campaign, the Jane Coffin Childs Memorial Fund for Medical Research, the Anna Fuller Fund, and the National Cancer Institute, U.S. Public Health Service.

References

[1] LOVELESS, A., and STOCK, J. C., Proc. Roy. Soc. B., in press.
[2] LOVELESS, A., and STOCK. J. C., Proc. Roy. Soc. B., in press.
[3] MARSHALL, R. D., Thesis for Ph. D., London (1955).
[4] LOVELESS, A., and STOCK, J. C., 1959: Proc. Roy. Soc. B. **150**, 423.
[5] LOVELESS, A., and STOCK, J. C., 1959: Proc. Roy. Soc. B. **150**, 486.
[6] LOVELESS, A., 1959: Proc. Roy. Soc. B. **150**, 497.
[7] HERSHEY, A. D., 1955: Virology **1**, 108
[8] REVELL, S. H., 1953: Heredity (Suppl.) **6**, 107.
[9] LITMAN, R. M., and PARDEE, A. B., 1956: Nature **178**, 529.
[10] LOVELESS, A., 1958: Nature **181**, 1212.
[11] VIELMETTER, W., and WIEDER, C. M., 1959: Z. Naturforsch. **14**, 312.
[12] RAPPOPORT, I. A., 1947: Dok. Vsesoyuz. Akad. Sel'sko-Khoz Nauk V. I. Lenina **12**, 12.
[13] FAHMY, O. G., and FAHMY, M. J., 1957: Nature **180**, 31.
[14] KØLMARCK, G., 1956: C. R. Lab. Carlsberg, Sér. physiol. **26**, 205.
[15] HESLOT, H., and FERRARY, R., 1958: Ann. Inst. Nat. Agronomique **44**, 3.
[16] HESLOT, H., FERRARY, R., LEVY, R., and MONARD, C., 1959: C. R. Acad. Sci. **248**, 729.

Inaktivierende und mutagene Wirkung der chemischen Veränderung von Nucleotiden in Virus-Nucleinsäure[1])

H. SCHUSTER, A. GIERER und K. W. MUNDRY

Max-Planck-Institut für Virusforschung und Max-Planck-Institut für Biologie, Abt. Melchers, Tübingen

1. Einleitung

Zur Untersuchung der Struktur und Funktion von Ribosenucleinsäure ist das Tabakmosaikvirus ein besonders geeignetes Modell. Ein Viruspartikel besteht aus einer Ribosenucleinsäure-Komponente von ca. 6000 Nucleotiden und einer Eiweißkomponente von 2200 gleichen Untereinheiten von je 160 Aminosäuren (Abb. 1). Die isolierte Nucleinsäure besitzt die Infektionseigenschaften des Virus und ist somit Träger der Erbfaktoren des TMV [1]. Sie besteht aus einem einzelnen Strang von Nucleotiden, die Integrität des Stranges ist Voraussetzung der Infektiosität [2, 3, 4, 5]. Die genetischen Eigenschaften des Virus beruhen somit auf der linearen Anordnung von ca. 6000 Nucleotiden im Nucleinsäurestrang.

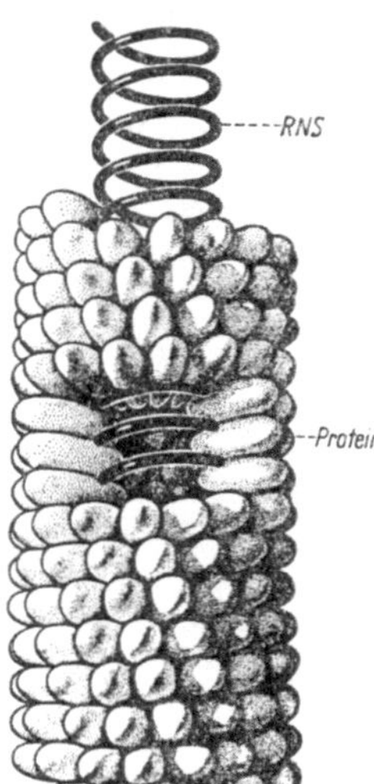

Abb. 1. Modell des Tabakmosaikvirus nach FRANKLIN und KLUG (1958) [16].

Die Ergebnisse führen zu der Frage, welche Bedeutung das einzelne Nucleotid für die Infektiosität und die Eigenschaften des Virus besitzt. Durch direkte chemische Analyse der RNS kann diese Frage derzeit nicht geklärt werden. Jedoch ist es möglich, Nucleotide in der RNS chemisch zu verändern und durch die kinetische Analyse auf die biologische Wirkung der Veränderung einzelner Nucleotide zu schließen.

Von besonderem Interesse sind solche Veränderungen, bei denen ein Nucleotid in ein anderes der vier die RNS aufbauenden Nucleotidarten – Adenin, Guanin, Cytosin und Uracil – oder in eine diesen Nucleotidarten analoge Base verwandelt wird. Eine hierfür sehr geeignete Reaktion ist die chemische Desaminierung durch salpetrige Säure. Hierbei erfolgen drei Arten von Umwandlungen (Abb. 2).

Im folgenden wird über zwei Arbeiten berichtet, die kürzlich über die biologische Wirkung dieser Desaminierungsreaktionen erschienen sind: In der ersten Arbeit wurde gezeigt, daß

[1]) Zur Diskussion gestellt von G. MELCHERS, Tübingen.

die Desaminierung eines einzelnen Nucleotids letal sein kann, und daß zumindest die Mehrzahl der 6000 Nucleotide von spezifischer Bedeutung für die Infektiosität ist [6]. Im Anschluß an dieses Ergebnis und die zugrunde liegenden chemischen Untersuchungen wurde gezeigt, daß die chemische Desaminierung jeweils einer einzelnen Base die genetischen Eigenschaften des Virus verändern kann [7, 8, 9]. Im Gegensatz zu den früher untersuchten chemischen und physikalischen Einwirkungen gestattet also die Desaminierungsreaktion den Nachweis einer Mutationsauslösung durch chemische Veränderung molekularer Verbindungen in vitro.

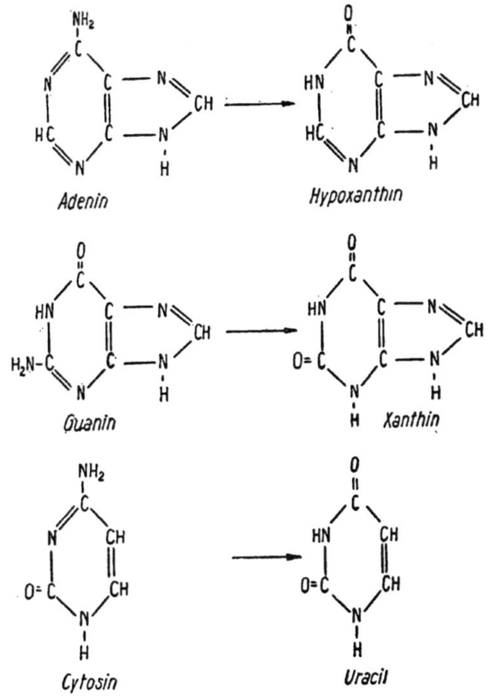

Abb. 2. Desaminierung von Nucleotiden durch salpetrige Säure.

2. *Inaktivierende Wirkung der Desaminierung*

Läßt man infektiöse RNS in schwach saurer Lösung mit HNO_2 reagieren, so werden nach Abb. 1 die Aminobasen des RNS-Moleküls in die entsprechenden Oxybasen umgewandelt. Außer dieser Desaminierung sind keine Nebenreaktionen der salpetrigen Säure mit der RNS nachzuweisen; das Molgewicht der RNS bleibt bei der Reaktion erhalten, es erfolgen somit keine Brüche in der Polynucleotidkette.

Um feststellen zu können, wie viele der mehreren Tausend Aminobasen in einem RNS-Molekül desaminiert werden dürfen, ohne daß die Infektiosität des Moleküls verlorengeht, wurde folgender Versuch durchgeführt: TMV-RNS wurde bei pH 4,3 mit HNO_2 umgesetzt, aus dem Ansatz nach verschiedenen Zeiten Proben entnommen und die Konzentration an infektiösen Molekülen im Pflanzentest bestimmt. Als Testpflanze wurde *Nicotiana glutinosa* verwendet, auf deren Blättern TMV und TMV-RNS lokale Nekrosen hervorrufen, deren Zahl ein Maß für die Konzentration an infektiösem Material ist.

Abb. 3 zeigt, daß bei HNO_2-Einwirkung die Infektiosität (Lokalläsionenzahl) exponentiell mit der Zeit abnimmt.

Um ausschließen zu können, daß durch die Reaktionsbedingungen allein (saures pH, hohe Konzentration an Salz) schon eine Inaktivierung der RNS erfolgen kann, wurde in einem Kontrollexperiment die RNS dem gleichen pH und der gleichen Salzkonzentration (an Stelle von $NaNO_2$ wurde Na-Acetat verwendet) wie beim HNO_2-Versuch ausgesetzt. Aus Abb. 3a kann man ersehen, daß die Kontroll-RNS nicht inaktiviert wird.

In dem Zeitraum, in dem die Infektiosität der RNS durch HNO_2 zerstört wird, ist die Zahl der desaminierten Nucleotide so gering, daß deren chemische Bestimmung noch nicht möglich

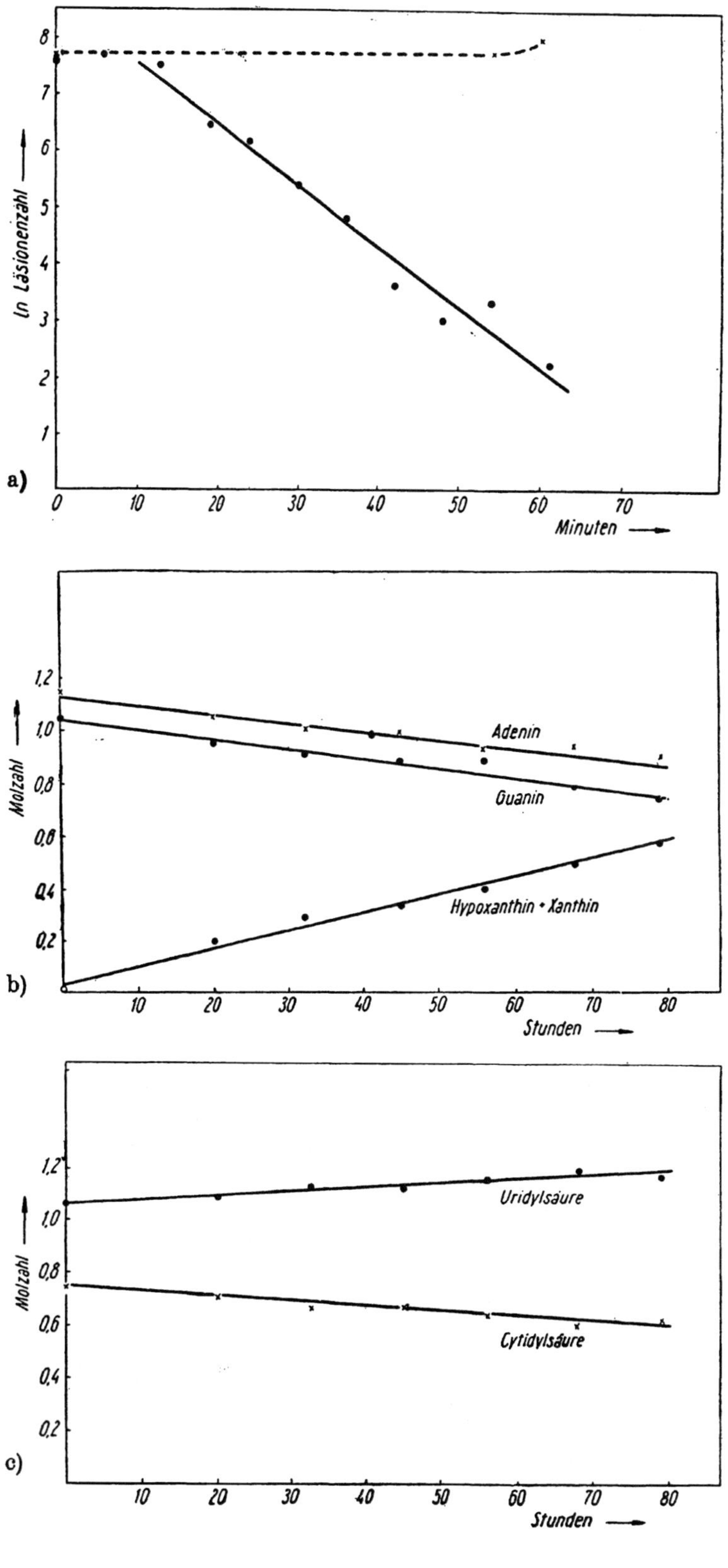

Abb. 3. Behandlung der TMV-RNS mit HNO_2 bei pH 4,3 (zit. nach [6]).

a) Inaktivierungsgeschwindigkeit der RNS, Läsionenzahl in Abhängigkeit von der Reaktionszeit .—.— RNS + HNO_2, x---x RNS + Kontrolle, (Anm.: Der exponentielle Abfall der Läsionenzahl erfolgt nur scheinbar nicht von der Zeit $t = 0$ ab, weil wegen der hohen Läsionenzahl am Anfang die Proportionalität zur Konzentration der infektiösen RNS nicht mehr gewahrt ist).

b) Molverhältnisse der Purine in Abhängigkeit von der Reaktionszeit mit HNO_2,

c) Molverhältnisse der Pyrimidine in Abhängigkeit von der Reaktionszeit mit HNO_2.

ist. Deshalb wurde die Behandlung mit HNO_2 unter konstanten Reaktionsbedingungen fortgesetzt und in längeren Zeitabständen Proben entnommen und eine Basen- bzw. Nucleotidbestimmung durchgeführt. Die graphische Darstellung (Abb. 3b und c) der Analyse läßt erkennen, daß die Menge an Aminobasen linear mit der Reaktionszeit abnimmt und die Menge an Oxybasen im gleichen Maße zunimmt. In Tab. 1 sind die Reaktionsgeschwindigkeiten α der Aminobasen angegeben. Da die Anteile dieser Basen in der RNS nicht gleich groß sind, ist zu erwarten, daß deren Reaktionsgeschwindigkeiten auch verschieden sind. Dagegen sind die auf gleiche Molanteile bezogenen Reaktionsgeschwindigkeiten α/m von gleicher Größenordnung.

Der exponentielle Abfall der Infektiosität bei HNO_2-Einwirkung entspricht einem Eintreffervorgang, die Inaktivierung der RNS erfolgt damit jeweils durch die Desaminierung („Treffer") eines einzelnen Nucleotids. Führt im Mittel die Desaminierung von einem unter insgesamt N Nucleotiden in der RNS zur Inaktivierung, so folgt die Abnahme der Infektiosität im Lauf der Zeit der Gleichung

$$J_t = J_0 \, e^{-N\alpha t}$$

bzw.

$$\ln J_0 - \ln J_t = N\alpha t .$$

J_t (Infektiosität zur Zeit t), J_0 (Infektiosität zur Zeit $t = 0$), α (Wahrscheinlichkeit der chemischen Veränderung eines beliebigen Nucleotids in der Zeiteinheit) und t (Reaktionszeit) werden experimentell bestimmt. Die Berechnung von N ergibt einen Wert von ungefähr 3000. Demnach führt die Desaminierung eines von etwa 3000 Nucleotiden in der RNS zur Inaktivierung. Da die Gesamtzahl der Nucleotide etwa 6000, also das Doppelte des für N berechneten Wertes beträgt, wirkt eine Desaminierung einer Base mit 50 % Wahrscheinlichkeit inaktivierend. Desaminierungen, die keinen Verlust der Infektiosität zur Folge haben, sind daher möglich. Sie könnten mutagen wirken.

Welche der nach Abb. 2 möglichen Desaminierungen letal wirken, kann in diesem Versuch nicht ermittelt werden, da die Reaktionsgeschwindigkeiten der einzelnen Aminobasen nach Tab. 1 von gleicher Größenordnung sind, somit die Chance, desaminiert zu werden, für alle Aminobasen im Molekül gleich groß ist.

Tabelle 1. (zit. nach [6]). Reaktionsgeschwindigkeiten α der RNS-Basen bei der Behandlung der TMV RNS mit HNO_2
α = umgesetzte Mole/Min. bezogen auf 4 Mol. Basen

Basen	Molverhältnisse m in der TMV-RNS (Summe = 4)	Versuch bei pH 4,3	
		$\alpha \cdot 10^5$	$\alpha/m \cdot 10^5$
Adenin	1,17	5,6	4,8
Guanin	1,03	6,2	6,0
Adenin + Guanin*)	2,20	11,5	5,2
Cytosin	0,73	3,0	4,1

*) Die Werte sind der Hypoxanthin + Xanthin-Kurve entnommen.

3. *Mutagene Wirkung der Desaminierung*

Die mutagene Wirkung der Desaminierung von RNS wurde an Hand eines Infektionssymptoms nachgewiesen, das für quantitative Untersuchungen besonders gut geeignet ist, nämlich der Erzeugung von Lokalläsionen auf Java-Tabak. Auf dieser Tabakrasse ruft der

Virusstamm *vulgare* nur chlorotische Läsionen hervor, die zu systemischen Infektionen führen, während Virusmutanten, die dem Ausgangsstamm nahe verwandt sind, nekrotische Lokalläsionen erzeugen [10, 11]. Es ergab sich, daß auch die Behandlung von Virus-RNS des normalen Stammes mit salpetriger Säure zu einer großen Anzahl nekrotischer Lokalläsionen führt.

Es war nun nachzuweisen, daß derartige Mutationen auf die Desaminierung einzelner Nucleotide der RNS *in vitro* zurückzuführen sind. Dieser Nachweis erfolgte in drei Schritten: Zunächst wurde gezeigt, daß die nekrotischen Lokalläsionen zum weitaus größten Teil auf neu entstandene Mutanten und nicht auf Selektion bereits vorhandener nekrotisierender Varianten zurückzuführen sind; sodann wurde nachgewiesen, daß diese Mutanten durch chemische Reaktion der Nucleinsäure entstanden sind. Schließlich ergab sich, daß die Desaminierung eines einzelnen Nucleotids mutagen sein kann.

a) Ausschluß der Selektion

Infiziert man Java-Pflanzen mit Virus-RNS des Stammes „vulgare" bei niedriger Infektiosität[1]), so erhält man etwa 2–3 nekrotische Läsionen auf 1000 Infektionsherde. Dieser Anteil kann auf einer Verunreinigung des Ausgangsmaterials mit nekrotisierenden Varianten, oder auf der Entstehung von Mutanten im Verlauf der Infektion beruhen. (Wahrscheinlicher ist das letztere, da auf Java-Tabak vermehrtes *vulgare* die gleiche Zahl von 2–3 primären Nekrosen je 1000 Infektionsherde auf Java-Blättern entwickelt). Im Durchschnitt aller Versuche mit relativ geringen Konzentrationen (ca. 10^{-6} g Virus/ml oder 2×10^{-5} g Virus-RNS/ml) ist diese „Spontanrate" 0,2‰. Im Einzelfall überschritt sie nie 0,4‰. – Trägt man RNS, die verschiedene Zeiten mit salpetriger Säure behandelt wurde, in gleichen Konzentrationen auf Javablätter auf, so findet man, daß die Zahl der Lokalläsionen mit der Dauer der Nitrit-Inkubation zunächst stark ansteigt, ein Maximum erreicht und schließlich infolge der inaktivierenden Wirkung der salpetrigen Säure wieder abfällt. (Abb. 4). Das Maximum

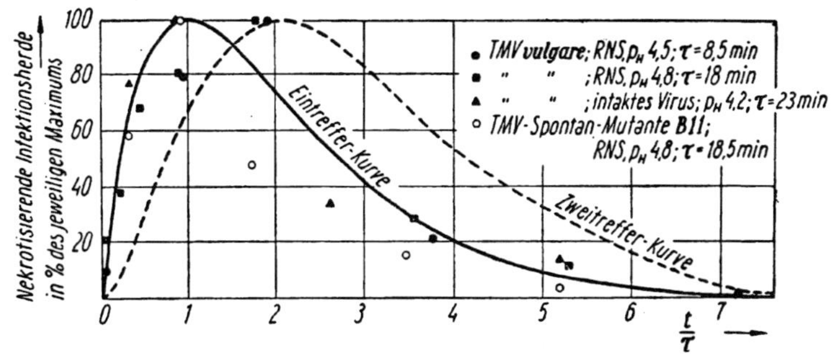

Abb. 4: Abhängigkeit der Zahl der Nekrosen pro Blatt auf Java-Tabak von der Dauer der Nitritinkubation. Die Zahl der Nekrosen pro Blatt ist in Prozenten des Maximalwertes angegeben, die Dauer der Nitritinkubation t ist auf die Zeit τ bezogen, nach der eine Inaktivierung um den Faktor e (2,7) erfolgt ist.

---

[1]) Zwischen „Infektiosität" und „Konzentration" muß streng unterschieden werden. Gegenüber den Konzentrationen in g Substanz/ml ausgedrückt, ist „Infektiosität" als Konzentration an infektiösen Partikeln in einem Präparat aufzufassen. Ein Maß dafür ist die Zahl an Infektionsherden je Flächeneinheit, z. B. je Blatt. Eine gegebene Viruslösung kann durch Inaktivierung der Partikel, aber auch durch deren Aggregation, eine viel geringere Infektiosität haben als eine gleich konzentrierte, aber unbehandelte mit normaler (Mono-)Dispersion der Partikel.

der Lokalläsionen ist erreicht, wenn die Infektiosität der RNS auf etwa 37 % des Ausgangswertes, d. h. um einen Faktor 2,7 gesunken ist. Die Zahl der Lokalläsionen/Blatt beträgt bei dieser Nitriteinwirkung das 20fache des Spontanwertes und das 50fache einer Kontrolle gleicher Infektiosität (Tab. 2). Dieses Ergebnis kann nicht durch eine Selektion von Mutanten erklärt werden, die den Ausgangsstamm verunreinigen. Selektion von gegen salpetrige

Tabelle 2. Absolute und relative Zunahme nekrotischer Läsionen, die von dem nicht-nekrotischen, aber HNO_2-inkubierten Stamm *vulgare* auf Javablättern hervorgerufen werden.

RNS 10^{-6} g/ml	Inkubation mit 1 m HNO_2 bei pH 4,8	Gesamtinfektiv. chlorot. u. nekr. Läsionen/Blatt	nekrot. Läsion/ Blatt	Rate: nekr. Läs./ Totalinfekt.
19	0	183	0,3	0,21
19	16	97	5,4	5,6
19	32	63	6,6	10,5
19	0	42	0,13	0,31

Säure resistenten Mutanten kann bei konstanter Konzentration des Inoculums und Ausschluß von Interferenz (s. u.) zu keiner absoluten Erhöhung der Anzahl ihrer Infektionsherde führen. (Bei fallender Infektiosität dagegen würde sie zu einer Abhängigkeit der „Mutationsrate" von der Inaktivierung führen, die mit den experimentellen Befunden unvereinbar ist [Abb. 5].) Selektion durch Interferenz („mutual exclusion") der Stämme hängt von der Infektiosität der Präparate ab, d. h. vom Abstand der Infektionsherde auf dem Blatt [12].

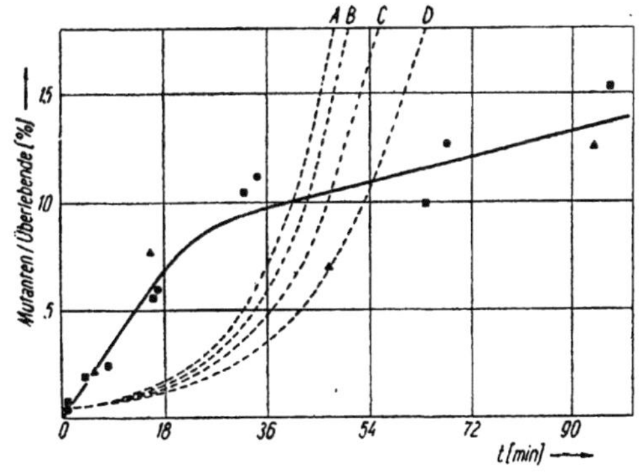

Abb. 5. Die Erzeugung von Mutanten durch HNO_2 oder die Selektion von gegen HNO_2 resistenten Varianten?

Abszisse: Zeit der Inkubation mit HNO_2 in Vielfachem von τ. (τ = Zeit, die zur Inaktivierung auf 37% der Ausgangsaktivität von TMV oder TMV-RNS notwendig ist). Ordinate: nekrotische Läsionen auf Java-Tabak in % der infektiösen Zentren (= % von Mutanten pro Überlebende) —— Kurve der Mutationsrate (siehe Abb. 5 MUNDRY u. GIERER 1958 [7]).
(3 Experimente: ● TMV-RNS, 39×10^{-6} g/ml; ■ TMV-RNS 19×10^{-6} g/ml; ▲ intaktes TMV, $0{,}7 \times 10^{-6}$ g/ml). ---- = theoretische Kurven der Selektion „A" völlig gegen HNO_2 resistenter, „B" 20 × resistenter, „C" 5 × resistenter, „D" 3 × resistenter Varianten des TMV in Abhängigkeit von der Inkubationszeit. Die Berechnung dieser theoretischen Kurven wurde auf der Basis ungünstigster Annahmen für das Vorhandensein resistenter Varianten (0,4%) und Interferenz durchgeführt (s. MUNDRY 1959 im Druck [13]).

Sie kann unberücksichtigt bleiben, wenn die Infektiosität der Kontrollen durch Verdünnen auf die der inaktivierten Präparate reduziert worden ist. Eine Inaktivierung um den Faktor 2,7 kann die Zahl der Primärinfektionen der Mutanten infolge Nachlassens der Interferenz keinesfalls um mehr als den gleichen Faktor erhöhen. Tatsächlich liegt der experimentell an Hand von Mischungen verschiedener Stämme bestimmte Selektionsfaktor noch wesentlich niedriger. Die durch die Nitriteinwirkung hervorgerufene Erhöhung der Zahl der Primärinfektionen um den Faktor 20 gegenüber dem Ausgangswert beruht daher überwiegend auf der Neuproduktion von Mutanten.

Bei sehr niedrigen Infektiositäten scheiden Interferenzphänomene als Selektionsursache überhaupt aus (Abb. 6), es sei denn, daß eine Aggregation infektiöser Partikel Selektion veranlaßt. Diese Art von Selektion ist ausgeschlossen, da das Nucleoproteid des TMV und die isolierte RNS quantitativ das gleiche Ergebnis liefern, obwohl deren Aggregationseigenschaften sich verschieden auswirken müßten. Hinzu kommt, daß aggregationsbedingte Selektion nur zu einer Minderbestimmung der jeweils selteneren Form, hier also der Mutanten, führen müßte, weil die Mutanten infolge Interferenz mit der dominierenden Komponente in einem Aggregat gegenüber monodispersen Verhältnissen eine verminderte Ausprägungswahrscheinlichkeit ihrer charakteristischen Symptome haben.

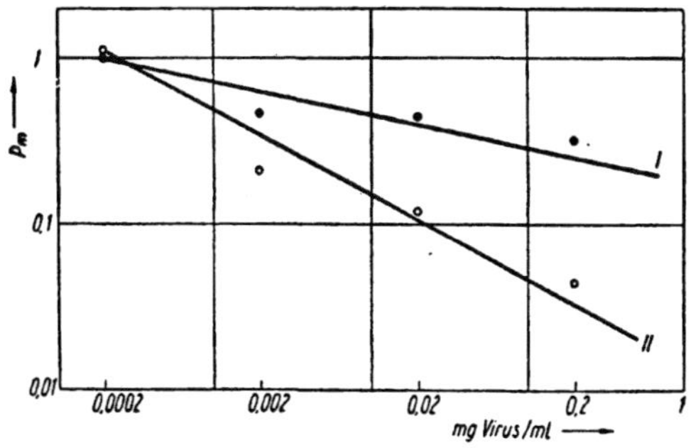

Abb. 6. Die Abhängigkeit der Selektion durch Interferenz zweier Virusstämme (nekrotisch und systemisch) von der absoluten Konzentration des Inoculums.

Abszisse: absolute Viruskonzentration.

Ordinate: Manifestationswahrscheinlichkeit $(p_m) = \frac{\text{Zahl nekrotischer Läsionen in Gegenwart nicht nekrotischer}}{\text{Zahl nekrotischer Läsionen in Abwesenheit nicht nekrotischer}}$.

I. und II. zwei verschiedene Versuche aus MUNDRY (1957) [12] siehe auch MUNDRY im Druck [13]. [12].

Damit gibt es keine Möglichkeit, das Auftreten nekrotisierender Infektionsherde infolge der Einwirkung von HNO_2 auf das (nicht nekrotisierende) Ausgangsmaterial irgendwie durch Selektion zu erklären [13]. Einer diesbezüglichen Kritik [14] fehlt daher die Grundlage. Es muß sich vielmehr um Virusmutanten handeln, die infolge einer unmittelbaren Reaktion zwischen dem Mutagen (der HNO_2) und dem Virus in vitro entstanden sind.

b) Ausschließung der Beteiligung von Protein

Die Mutanten entstehen sowohl bei der Behandlung isolierter Nucleinsäure als auch von intaktem Virus mit salpetriger Säure. Die gemessene Inaktivierung läßt sich hierbei jeweils als ein Maß für die chemische Reaktion der RNS verwenden. Es ergibt sich, daß bei gleicher Inaktivierung, also bei gleicher Reaktion der RNS, jeweils der gleiche Prozentsatz von Lokalläsionen unter den gesamten Primärinfekten auftritt. Wird TMV mit Nitrit behandelt, so ergibt sich derselbe Prozentsatz Mutationen, wenn das TMV als solches und wenn die nach der Nitritbehandlung aus ihm extrahierte RNS aufgetragen wird. Hieraus wird geschlossen, daß die Mutationen ausschließlich auf chemischen Reaktionen an der RNS beruhen (vgl. Abb. 4).

c) Mutagene Wirkung der Desaminierung einzelner Basen

Die mutagene Wirkung der Änderung einzelner Basen ergibt sich aus der Kinetik der Erzeugung von Mutanten durch Desaminierung. Nur in diesem Falle ist zu Beginn der Reaktion ein linearer Anstieg der Mutanten mit der Inkubationsdauer zu erwarten. Wenn mehrere Nucleotide verändert werden müßten, um eine Mutation hervorzurufen, so wäre ein parabolischer Anstieg zu erwarten. Die Messungen des Anstiegs bestätigen quantitativ, daß Mutanten durch Desaminierung einzelner Basen entstehen.

Die diskutierten Ergebnisse sind in Abb. 4 zusammengefaßt. Aufgetragen ist die Zahl der Lokalläsionen, die von einer konstanten Konzentration von Virus bzw. RNS hervorgerufen wird (bezogen auf den Maximalwert) gegen die Zeit der Nitritinkubation. Die Inkubationsdauer ist hierbei auf diejenige Zeit bezogen, innerhalb der die Gesamtinfektivität jeweils auf 37 % des Ausgangswertes sinkt. Diese Zeitskala mißt direkt die Zahl der desaminierten Nucleotide [6]. Die eingetragene, ausgezogene Kurve stellt die theoretische Relation für den Fall dar, daß die Desaminierung einzelner Nucleotide mutagen ist. Die Meßreihen, die sowohl mit intaktem TMV als auch mit isolierter RNS unter verschiedenen Bedingungen der Inkubation mit salpetriger Säure und bei verschiedenen Konzentrationen der Infektion ausgeführt sind, bestätigen alle die theoretisch erwartete Beziehung.

d) Das Mutationsspektrum

Das TMV *vulgare*[1]) ist für seine mannigfaltigen Mutationsmöglichkeiten zu zahlreichen Stämmen mit z. T. sehr unterschiedlichen Symptomen bekannt. Es sei hier nur an Gelbstämme, maskierte oder abgeschwächte Typen und nekrotisierende mit und ohne Sekundärinfektion erinnert. Es ist zu prüfen, ob nach Behandlung mit HNO_2 auch solche Mutanten mit deutlich erhöhter Rate auftreten.

Tabelle 3. Übersicht über die nach Prüfung auf Samsun-Tabak und Java-Tabak in nitritinkubierter vulgare-RNS gefundenen, vom unbehandelten Ausgangsmaterial abweichenden Typen (Versuchsdaten siehe Tab. 2)

Präparat = RNS aus *vulgare*-TMV	96 min mit Nitrit behandelt 20×10^{-6} g RNS/ml		Unbehandelte Verdünnungs-Kontrolle $0{,}19 \times 10^{-6}$ g RNS/-ml	
Testpflanzen	Samsun-Tabak[1]	Java-Tabak[1]	Samsun-Tabak	Java-Tabak
Bezeichnung der Symptom-Typen				
nicht angegangene Abimpfungen	20	17	1	1
wahrscheinlich unverändertes Virus	7	12	64	64
vulgare-ähnlich, doch ohne wesentliche Deformation der Blätter	5	2	0	0
Typen mit deutlicher Adernbänderung	7	5	0	0
± maskierte Stämme	2	0	0	0
langsam sich ausbreitende Mosaiktypen	3	6	0	0
abgeschwächte Typen	2	2	0	0
Stämme mit verstärkten (bis schweren) Blatt-Deformationen	3	2	0	0
hellgrüne Mosaikstämme	4	1	0	0
systemisch nekrotisierende Stämme	2	0	0	0
atypische Einzelformen	3	1	0	0
nur mit Primärinfektionen:				
a) nekrotisierend	1	4	0	0
b) chlorotisierend	1	2	0	0
c) schwach chlorotisierend	0	5	0	0
d) unsicher, ob mit (Primär-) Infektion	0	1	0	0
Summe	60	60	65	65
Durchgeführte Einzelherd-Abimpfungen	60	60	65	65

[1]) Keine nekrotische Reaktion auf Java-Tabak oder *Nicotiana silvestris*, sondern systemische mit hellgrün-dunkelgrün Mosaik und nur ± leichten Blattdeformationen.

Auf *Nicotiana tabacum* var. „*Xanthi*" erzeugen alle Stämme des TMV unabhängig von etwaigen sonstigen Symptomen auf anderen Testpflanzen lokale Nekrosen, ebenso wie das bei *N. glutinosa* der Fall ist. Systemische Infektion tritt nicht ein. Diese Nekrosen gehen auf Infektion mit einem einzelnen TMV-Partikel bzw. infektiösen RNS-Molekül zurück. In Übertragungen solcher Einzelherde auf türkischen Tabak (Samsun-Tabak) und jeweils parallel dazu auf Java-Tabak besteht eine Möglichkeit, mit der man ein Viruspräparat auf seine Reinheit hin untersuchen kann. Mit diesem Verfahren wurde nun inkubiertes Virus mit einer Verdünnungskontrolle verglichen, die die gleiche Zahl von Nekrosen je Xanthi-Blatt erbracht hatte. Von beiden Gruppen wurden alle entstandenen Infektionsherde isoliert und getestet.

Die Ergebnisse sind in Tab. 3 zusammengefaßt. Sie zeigen, daß mehrmals soviel Java nicht nekrotisierende als nekrotisierende Mutanten auftreten, und auch, daß HNO_2 keine spezifische mutagene Wirkung hat. Die Wirkung von HNO_2 ist also nicht auf den Merkmalsunterschied systemisch-nekrotisch beschränkt.

Diskussion

Sowohl die inaktivierende wie auch die mutagene Wirkung der Desaminierung einzelner Basen der RNS bestätigt die Auffassung, daß die genetischen Eigenschaften des Virus durch die Aufeinanderfolge der Nucleotide im RNS-Strang bestimmt sind. Verändert man jeweils ein einzelnes Nucleotid, so hat dies in etwa der Hälfte der Fälle inaktivierende Wirkung. Zumindest ein Teil der übrigen Desaminierungen ist mutagen. Hierbei ist experimentell nicht geklärt, welche der drei möglichen Desaminierungen „letal" bzw. mutagen sein können.

Bei einer Umwandlung von Cytosin in Uracil ist es möglich, daß derartige RNS identisch redupliziert wird und daher zu biologisch noch wirksamem, aber genetisch verändertem Virusmaterial führt. Bei der Umwandlung von Adenin in Hypoxanthin entsteht zwar eine Base, die in RNS nicht vorkommt, jedoch ist diese dem natürlichen Baustein Guanin sehr nahe verwandt. Es ist bekannt, daß in DNS tatsächlich Hypoxanthin an Stelle von Guanin

Tabelle 4. Reaktion von TMV-RNS mit $(CH_3)_2SO_4$. Inaktivierung und Mutation (nach SCHUSTER und MUNDRY 1958)

relative Virus-konzentration	Inkubations-dauer (min)	Infektiosität auf Xanthi-Tabak (15 Blätter)	Nekrosen auf Javatabak (32 Blätter)	Rate Java-Nekr. / Xanthi-Nekr.	Mittelwert dieser Rate
1	0	6221	19	$3{,}06 \times 10^{-3}$	$3{,}44 \times 10^{-3}$
0,1	0	1568	6	$3{,}82 \times 10^{-3}$	
0,01	0	120	0	—	
0,001	0	21	0	—	
[1][1])	0	5982	16	$2{,}68 \times 10^{-3}$	$3{,}26 \times 10^{-3}$
[0,18]	3	2220	5	$2{,}26 \times 10^{-3}$	
[0,01]	6	1365	3	$2{,}20 \times 10^{-3}$	
[0,034]	11	513	3	$5{,}84 \times 10^{-3}$	
[0,01]	16	116	0	—	

[1]) = Werte in [ ] wurden aus den Infektionsdaten über die aus den Kontrollen ermittelten Konzentrations-Infektiositätskurven ermittelt.

eingebaut werden kann. Es ist demnach denkbar, daß die Entstehung von Hypoxanthin in RNS ebenfalls zu Mutanten führt, wobei in den Nachkommen Guanin an Stelle eines Adenin zu finden wäre. Die Wirkung der Desaminierung von Guanin zu Xanthin ist theoretisch schwer vorherzusagen, wahrscheinlich führt sie nur zu inaktiver Nucleinsäure.

Die hier beschriebene Möglichkeit, genetisches Material in vitro zu verändern, ist für das Studium des Zusammenhangs zwischen Nucleinsäure und Eiweißstruktur von Bedeutung. Insbesondere kann untersucht werden, inwieweit Veränderungen einzelner Nucleotide der RNS zu Veränderungen der Struktur des entsprechenden Virusproteins führt und welcher Art diese Veränderungen sind.

Von besonderem Interesse ist ferner, welche weiteren chemischen Veränderungen der Nucleinsäure in vitro mutagen sein können. Unveröffentlichte Versuche [15] ergaben, daß beispielsweise die Methylierung von Nucleotiden der TMV-RNS mit Dimethylsulfat zwar inaktivierend wirkt, daß sich aber Mutanten, die Lokalnekrosen auf Javatabak erzeugen, nicht nachweisen ließen (Tab. 4).

Literatur

[1] Gierer, A., und Schramm, G., 1956: Die Infektiosität der Nucleinsäure aus Tabakmosaikvirus. Z. f. Naturforsch. **11** b, 138.

[1a] Gierer, A., and Schramm, G., 1956: Infectivity of ribonucleic acid from tobacco mosaic virus. Nature **177**, 702.

[2] Gierer, A., 1957: Structure and biological function of ribonucleic acid from tobacco mosaic virus. Nature **179**, 1297.

[3] Gierer, A., 1958: Größe und Struktur der Ribosenucleinsäure des Tabakmosaikvirus. Z. f. Naturforsch. **13** b, 477.

[4] Gierer, A., 1958: Die Größe der biologisch aktiven Einheit der Ribosenucleinsäure des Tabakmosaikvirus. Z. f. Naturforsch. **13** b, 485.

[5] Mundry, K. W., 1958: Über die Korrelation zwischen Partikellänge und Infektiosität beim Vergilbungsvirus der Rüben. Z. f. Naturforsch. **13** b, 19.

[6] Schuster, H., und Schramm, G., 1958: Bestimmung der biologisch wirksamen Einheit des Tabakmosaikvirus auf chemischem Wege. Z. f. Naturforsch. **13** b, 697.

[7] Mundry, K. W., und Gierer, A., 1958: Der mutagene Effekt von salpetriger Säure auf Ribosenucleinsäure des Tabakmosaikvirus in vitro. Proc. Intern. Congr. Biochemistry, Wien.

[8] Gierer, A., and Mundry, K. W., 1958: Production of mutants of tobacco mosaic virus by chemical alteration of its ribonucleic acid in vitro. Nature **182**, 1457.

[9] Mundry, K. W., und Gierer, A., 1958: Die Erzeugung von Mutationen des Tabakmosaikvirus durch chemische Behandlung seiner Nucleinsäure in vitro. Z. f. Vererbungsl. **89**, 614.

[10] Melchers, G., 1942: Über einige Mutationen des Tabakmosaikvirus und eine „Parallelmutation" des Tomatenmosaikvirus. Naturwiss. **30**, 48.

[11] Melchers, G., Schramm, G., Trurnit, H., und Friedrich-Freksa, H., 1942: Die biologische, chemische und elektronenmikroskopische Untersuchung eines Mosaikvirus aus Tomaten. Biol. Zbl. **60**, 524.

[12] Mundry, K. W., 1957: Zur Frage des Einflusses von Röntgen- und UV-Strahlen auf die Mutationsrate des Tabakmosaikvirus nach Bestrahlung reiner Präparate. Z. f. Vererbungsl. **88**, 115.

[13] Mundry, K. W., 1959: The effect of nitrous acid on tobacco mosaic virus: mutation, not selection. Im Druck.

[14] Bawden, F. C., 1959: Effect of nitrous acid on tobacco mosaic virus: mutation or selection? Nature **184**, p. „B. A." 27.

[15] Schuster, H., und Mundry, K. W., 1958: unveröffentlicht.

[16] Franklin, R. E., and Klug, A., 1958: zitiert nach M. E. Perutz: Fortschritte in der Molekularbiologie. Endeavour **17**, 190.

Inaktivierung und Mutagenese von *Escherichia coli* durch Inkubation mit Natriumnitrit [1])

F. Kaudewitz

Max-Planck-Institut für Virusforschung, Tübingen

Werden Zellen von *Eschericha coli* mit Natriumnitrit in geeignetem pH inkubiert, so findet eine Inaktivation der Befähigung der Zellen zur Koloniebildung statt [1,2]. Das Ausmaß der Inaktivierung pro Zeiteinheit hängt dabei von der Nitritkonzentration und dem pH-Wert des Inkubationsgemisches sowie dem Stoffwechselzustand der behandelten Zellen ab. Diese

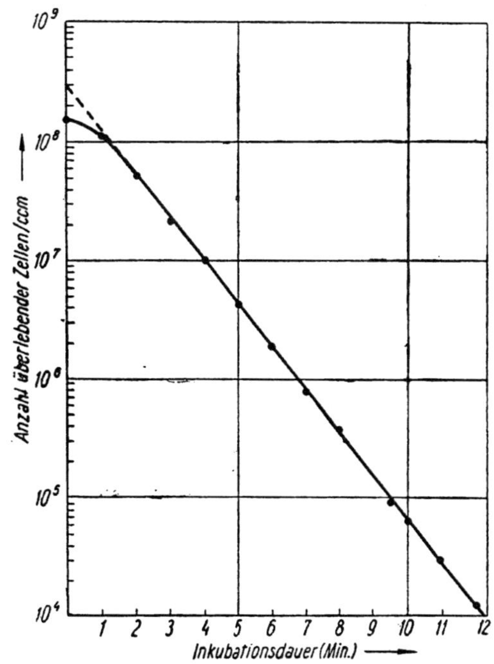

Abb. 1. Inaktivierung der Zellen einer sich in Stoffwechselruhe befindlichen Population von *E. coli* durch Inkubation mit 0,0167-m. Natriumnitrit, pH 4,5; bei 37° C = Standard-Inkubation.

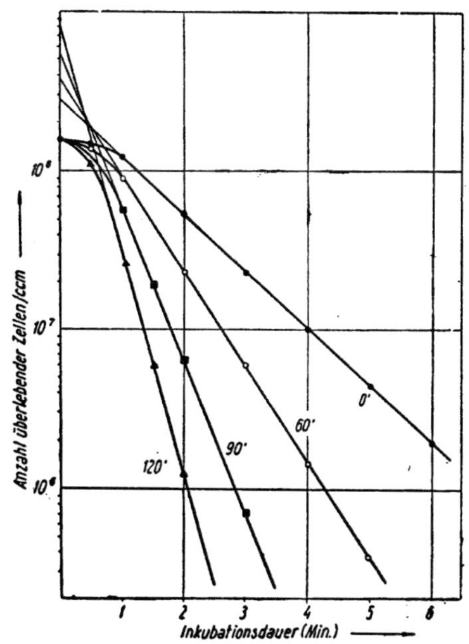

Abb. 2. Inaktivations-Kurven von *E. coli*-Zellen unterschiedlich intensiven Stoffwechsels. In Stoffwechselruhe befindliche Zell-Populationen wurden vor Inkubationsbeginn 0, 60, 90 und 120 Minuten lang in Bouillon unter Belüftung bebrütet.

[1]) Zur Diskussion gestellt von G. Melchers, Tübingen.

Inaktivationsrate nimmt für den geprüften Bereich von pH 5,8–4,1 mit steigender Wasserstoffionenkonzentration des Inkubationsgemisches zu. Gleichsinnig wächst dabei die Konzentration der durch Hydrolyse von $NaNO_2$ gebildeten salpetrigen Säure. Ein weiterer Weg diese Konzentration bei konstant gehaltenem pH zu vergrößern ist die Vermehrung des $NaNO_2$-Gehaltes. Auch dadurch wird die Inaktivationsrate erhöht. Aus all diesen Zusammenhängen darf geschlossen werden, daß undissoziierte salpetrige Säure das reagierende Agens der Inaktivation bildet.

Versuche mit ausgehungerten Zellen, welche keine nennenswerte Stoffwechselaktivität mehr aufweisen, ergeben die in Abb. 1 dargestellte Inaktivationskurve. Werden solche Zellen jedoch durch verschieden lange Bebrütung in Bouillon aus dem Zustand der Stoffwechselruhe in den unterschiedlich intensiven Stoffwechsels überführt, so nimmt mit steigender Stoffwechselintensität die Inaktivationsrate zu (Abb. 2). In Versuchen mit ruhenden, nicht in Stoffwechsel befindlichen Zellen ergibt die Abtragung des Logarithmus der Anzahl überlebender Zellen nach Inaktivation mit Nitrit gegen die Zeitdauer der Nitriteinwirkung eine Kurve, deren Grenzgerade für große t die Ordinatenachse beim doppelten Ausgangswert schneidet. Die Inaktivation solcher ruhenden Zellen durch $NaNO_2$ kann daher formal in folgender Weise als Zweitreffervorgang abgehandelt (genau genommen sind es zwei unabhängige Eintreffer-Vorgänge in zwei verschiedenen Treffvolumina) werden:

$$\frac{S}{S_0} = 1 - \left(1 - e^{-\frac{t}{\tau_1}}\right)^2 \qquad (1)$$

wobei S die Anzahl nicht inaktivierter Bakterien zur Zeit t, S_0 die Anzahl nicht inaktivierter Bakterien zur Zeit 0 und τ_1 die Zeitkonstante der Inaktivierung ist. Durch Erhöhung des pH-Wertes wird, wie oben dargestellt, die Reaktion verlangsamt, so daß der gekrümmte Kurventeil genauer ausgemessen werden kann (Abb. 3). Die Lage der Meßpunkte eines solchen Versuches stimmt mit dem Verlauf der Kurve gut überein, welche unter Verwendung der theoretischen Werte der Gleichung (1) gezeichnet wurde. Damit wird gezeigt, daß die Inaktivierung ruhender Zellen von *Escherichia coli* durch Nitrit ein Zweitreffervorgang ist.

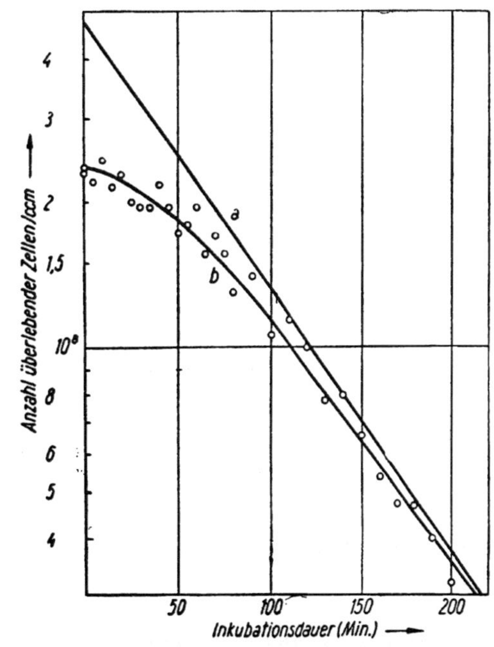

Abb. 3. Inaktivierung der Zellen von *E. coli* in Stoffwechselruhe bei pH 5,8 (37° C). a) Grenzgerade der Zweitreffer-Kurve. b) Verlauf der theoretischen Inaktivationskurve bei Vorliegen einer Zweitreffer-Kinetik des Inaktivationsvorganges.

Die Zweitrefferkinetik der Nitrit-Inaktivierung erlaubt den Schluß, daß zwei NH_2-Gruppen desaminiert werden müssen, um die Befähigung der Zelle zur Koloniebildung zu vernichten. Zu welchen Einheiten gehören diese NH_2-Gruppen? Arbeiten anderer Autoren sowie eigene Versuche [3, 4] lassen erkennen, daß in Stoffwechselruhe befindliche Zellen von *Escherichia coli* desjenigen Wachstumszustandes, wie sie zu den vorliegenden Untersuchungen benutzt wurden, einkernig sind. In einem einzigen Zellkern also müssen 2 NH_2-Gruppen desaminiert werden, um die Bakterienzelle zu inaktivieren. Unter den damit gegebenen Voraussetzungen wird die Zweitreffer-Kinetik der Inaktivierung als genetische Inaktivierung deutbar: Bakterielle DNS als chemisches Äquivalent der Gene dieser Zellen ist aus zwei Untereinheiten aufgebaut. Untersuchungen von MESELSEN und STAHL [5] beweisen, daß bei der identischen Duplikation jede dieser Untereinheiten verdoppelt wird, indem sie dabei gleichzeitig erhalten bleibt. Es ist naheliegend, in diesen Untereinheiten die beiden DNS-Stränge eines nach dem WATSON-CRICK-Modell [6] aufgebauten DNS-Moleküls zu sehen, die sich, verursacht durch die Paarung ihrer Basen, wie Positiv und Negativ verhalten. Wie Untersuchungen von GIERER und MUNDRY [7] zeigen, wird die aus einem Einzelfaden bestehende RNS des Tabakmosaikvirus durch die Desaminierung einer einzigen NH_2-Gruppe inaktiviert. Auf die DNS übertragen würde dies bedeuten, daß je eine NH_2-Gruppe in jedem der beiden Schwestermolekülfäden des WATSON-CRICK-Modelles desaminiert werden müßte, um zur Inaktivierung der Zelle zu führen. Wird dagegen nur eine der beiden Untereinheiten getroffen, so bleibt die genetische Information des Schwesterfadens – sei dieser in dem oben genannten Sinne das Positiv oder Negativ – unverändert. Die Duplikation eines solchen Fadens stellt den ursprünglichen Zustand des aus zwei intakten DNS-Fäden bestehenden WATSON-CRICK-Modelles wieder her: Ein einziger Treffer führt damit nicht zur Inaktivation. Das Vorliegen einer Zweitrefferkinetik der Inaktivation einkerniger Zellen von *Escherchia coli* macht es damit sehr wahrscheinlich, daß diese Inaktivierung durch Desaminierung zweier, den beiden Untereinheiten der DNS angehörenden Basen hervorgerufen wird.

Die Behandlung von Bakterienzellen mit Nitrit führt nicht nur zur Inaktivation. Sie wirkt gleichzeitig auch mutagen. Die mutagene Wirkung wurde qualitativ und quantitativ am Modellfall des Auftretens auxotropher Mutanten nach Inkubation von Wildtypzellen mit Nitrit untersucht. Kontrollen vermochten gleichzeitig den Einwand zu entkräften, daß ein in solchen Versuchen mit zunehmender Inkubationsdauer beobachtetes Ansteigen des Anteiles auxotropher Mutanten unter den überlebenden Zellen durch Selektion von schon vor Versuchsbeginn vorhandenen Mutanten hervorgerufen wird [1, 2]. Zur Durchführung derartiger Versuche wurden Zellen des Wildtyps von *Echerichia coli* B nach Nitritbehandlung auf Vollmedium plattiert und nach Bebrütung das Koloniemuster dieser Mutterplatten mit Hilfe der replica-plating Technik [8] auf Platten mit Minimalmedium übertragen. Die auf den Mutterplatten befindlichen Kolonien, deren Abdrücke auf Minimalmedium nicht weitergewachsen waren, wurden isoliert und erwiesen sich in etwa 80% aller Fälle als Mangelmutanten. Bei Anwendung der beschriebenen Versuchstechnik traten keine Kolonien auf, die aus Wildtyp- und auxotrophen Sektoren zusammengesetzt waren. Die Untersuchung der Abhängigkeit des prozentualen Anteiles auxotropher Mutanten unter den überlebenden Zellen von der Inkubationsdauer mit Nitrit ergab für ruhende nicht in Stoffwechsel befindliche Zellen eine Zweitreffer Kinetik. Sie läßt sich formal in folgender Weise abhandeln:

$$\frac{M}{S} = \left(\frac{t}{\tau_2}\right)^2, \tag{2}$$

solange $t \ll \tau_2$ ist.

Oder logarithmiert:

$$\log \frac{M}{S} = 2 \cdot \log \frac{t}{\tau_2} \tag{3}$$

(M = Anzahl der Mutanten zur Zeit t, τ_2 = Zeitkonstante der Mutationsauslösung).

Werden die erhaltenen Prozentwerte an Mutanten logarithmisch gegen den Logarithmus der Zeitdauer der Inkubation mit Natriumnitrit abgetragen (Abb. 4), so müßte nach Gleichung (3) bei Vorhandensein einer Zweitreffer-Kinetik der Auslösung auxotropher Mutanten die Verbindung der Beobachtungswerte eine Gerade mit der Steigung 2 sein. Die

Abbildung zeigt, daß diese Voraussagen zutreffen. Wurden zu diesen Versuchen jedoch in gleicher Weise, wie in Abb. 2 dargestellt, Zellen mit verschieden regem Stoffwechsel benutzt, so schnitt die Grenzgerade für große t der dabei gewonnenen Inaktivationskurve die Ordinate bei einem Mehrfachen des Ausgangswertes. Dennoch blieb stets trotz unterschiedlicher Inaktivationsrate die Abhängigkeit des prozentualen Anteiles auxotropher Mutanten unter den Überlebenden von der Zeitdauer der Inkubation unverändert (Abb. 5). Die Mutationsauslösung durch Nitritinkubation muß damit an die Desaminierung von Einheiten gebunden sein, welche unabhängig vom Stoffwechselzustand der Zelle bereits in ruhenden Zellen voll zur Verfügung stehen. Ein solcher Tatbestand macht es sehr wahrscheinlich, daß **die Mutationsauslösung durch direkte Reaktion salpetriger Säure mit den Basen der DNS hervorgerufen wird.** Indirekte mutagene Wirkung dagegen zeigt stets eine quantitative Abhängigkeit [9, 10, 11] vom Stoffwechselzustand der Zelle.

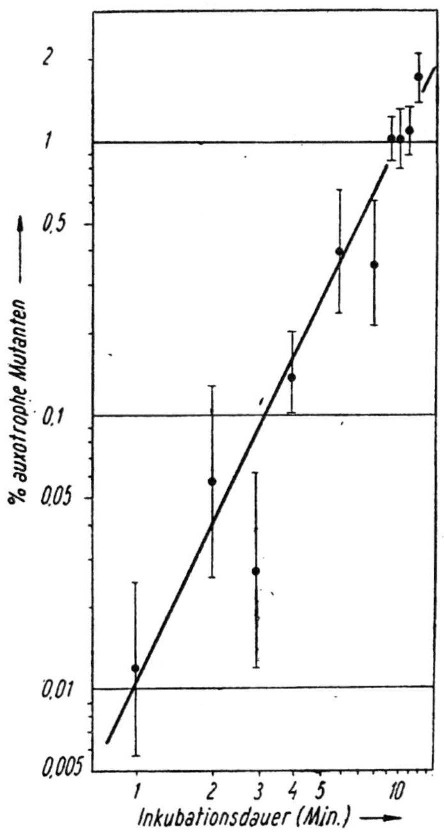

Abb. 4. Die Abhängigkeit des prozentualen Anteils auxotropher Mutanten an überlebenden, sich in Stoffwechselruhe befindlichen *E. coli*-Zellen von der Aufenthaltsdauer im Standard-Inkubationsgemisch. Eingezeichnete Mutungsgrenzen $= \pm 2\sigma$.

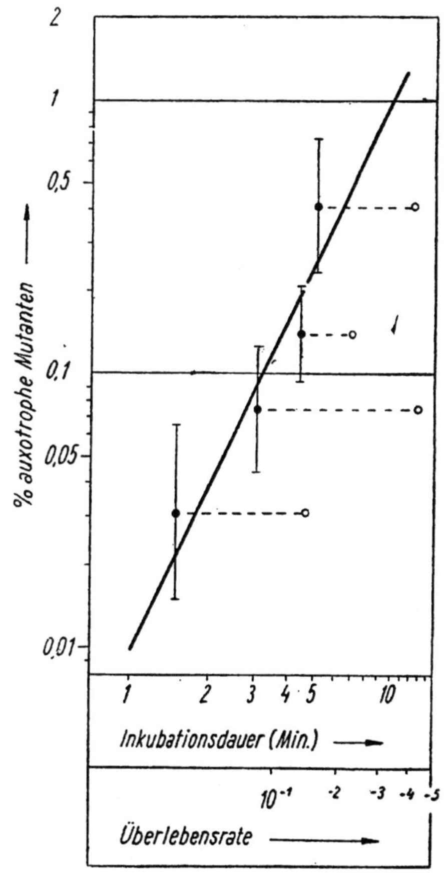

Abb. 5. Abhängigkeit des prozentualen Anteils auxotropher Mutanten an überlebenden, sich in unterschiedlich intensivem Stoffwechsel befindlichen *E. coli*-Zellen von der Inkubationsdauer im Standardgemisch (Punkte) und der Überlebensrate (Kreise). Eingezeichnete Mutungsgrenzen $= \pm 2\sigma$.

Die unter den dargestellten Versuchsbedingungen sich ergebende Zweitreffer-Kinetik der Mutationsauslösung läßt sich auf der Basis der dargestellten Zusammenhänge und der daraus gezogenen Schlüsse nicht deuten. Zur Entstehung einer mutierten Zelle müßte die Desaminierung einer einzigen geeigneten NH_2-Gruppe aus einer der beiden DNS-Untereinheiten ausreichen. Die diesem Ereignis folgende identische Duplikation der DNS würde dann zu einer mutierten Zelle und einer Wildtypzelle führen. Aus beiden entstünde auf Vollmedium eine gemischte Kolonie; die Mutationskinetik wäre eine Eintreffer-Kinetik. Eine solche Überlegung führt notwendigerweise zu der Frage, ob die im Versuch beobachtete Zweitreffer-Kinetik der Mutationsauslösung durch Nitritinkubation von *Escherichia coli* nicht durch die besondere Art der Versuchsdurchführung lediglich vorgetäuscht wird. Das wäre der Fall, wenn nur solche Mutanten zur Beobachtung gelangten, in deren Mutterzelle die eine der beiden DNS-Untereinheiten eine zur Mutation führende Desaminierung erfahren hätte, während die zweite Untereinheit durch eine weitere Desaminierung inaktiviert worden wäre. Dann dürften unter diesen Versuchsbedingungen im Einklange mit den experimentellen Befunden niemals sektorierte Kolonien auftreten. Die damit ausgesprochene Annahme würde folgende experimentell prüfbare Voraussetzungen enthalten: Wird innerhalb der Bakterienzelle nur eine der beiden Untereinheiten eines DNS-Moleküls mit mutagener Wirkung desaminiert, so weist diese Einheit oder die nach ihrer Duplikation sie beherbergende Zelle im Vergleich zu der unverändert gebliebenen Einheit oder der daraus hervorgehenden Wildtypzelle zunächst eine längere Zeitspanne zwischen zwei Teilungen auf. Die entstehende Kolonie enthält daher vorwiegend Wildtypzellen und nur einen geringen Prozentsatz auxotropher Zellen des gleichen Typs. Wird eine solche auf Vollmedium herangewachsene Kolonie auf Minimalmedium gestempelt, so überwachsen die in ihr enthaltenen Wildtypzellen die auxotrophen Mutanten. Die Kolonie scheint nur aus Wildtyp aufgebaut zu sein; sie wird als „nicht mutiert" gewertet. Durch die gleiche Versuchsanordnung könnten dann Mutanten nur nachgewiesen werden, wenn eine der beiden Untereinheiten durch Desaminierung mutiert, die andere durch einen zweiten Treffer inaktiviert wurde. Dadurch wäre die Beobachtung einer Zweitreffer-Kinetik als an die angewandten Versuchsbedingungen gebunden erklärt.

Neuere experimentelle Befunde stützen diese Voraussagen. Werden Zellen nach 10 Minuten langer Inkubation im Standard-Inkubationsgemisch [2] unmittelbar auf Vollmedium plattiert, so erweisen sich etwa 1 % der Überlebenden als auxotrophe Mutanten. Wird dagegen die gleiche Zellsuspension nach Inkubation in Bouillon eingebracht und bebrütet, so daß der Zelltiter auf das 10-fache ansteigt, dann enthält die Suspension weniger als 0,05 % Mutanten. Im Kontrollversuch läßt sich bei Verwendung nitritinduzierter, bereits über mehrere Passagen geführter, auxotropher Mutanten nachweisen, daß diese in Bouillon die gleiche Wachstumsrate wie Wildtypzellen aufweisen. Unmittelbar nach Induktion, und nur dann, ist daher im Einklange mit der oben gemachten Voraussetzung die Vermehrungsrate induzierter Mutanten wesentlich kleiner als die der Wildtypzellen. In einer weiteren Versuchsreihe wurden Wildtypzellen von *Escherichia coli* B unter Standardbedingungen mit Nitrit inkubiert, auf Vollmedium plattiert und einzelne der nach Bebrütung durch den Stempelversuch als dem Wildtyp angehörig ausgewiesenen Kolonien isoliert. Die aus ihnen gewonnenen Zellsuspensionen wurden nach geeigneter Verdünnung erneut auf Vollmedium plattiert und die nach Bebrütung entstandenen Kolonien auf Minimalmedium gestempelt. Dabei zeigte sich, daß mit wachsender Dauer der Nitritinkubation ein zunehmender Prozentsatz der aus überlebenden Zellen auf Vollmedium entstehenden Kolonien neben Wildtypzellen auch auxotrophe, innerhalb einer Kolonie stets dem gleichen Typ angehörenden Mutanten enthielt. In Kontrollen mit nicht nitritinkubierten Zellen konnten auf diesem Wege Mutanten nicht aufgefunden werden. Die experimentelle quantitative Prüfung dieser Beobachtungen wird zur Zeit durchgeführt. Alle derartigen bisher gewonnenen Ergebnisse stützen jedoch bereits heute die Annahme, daß die Mutationsauslösung durch Inkubation von Wildtypzellen der Art *Escherichia coli B* mit Nitrit einer Eintreffer-Kinetik folgt und die Zweitreffer-Kinetik (Abb. 4) durch die Art der Versuchsanordnung vorgetäuscht wird.

Literatur

[1] F. Kaudewitz, 1959: Nature **183**, 1829.
[2] F. Kaudewitz, 1959: Z. Naturf. **14**b, 528.
[3] E. Witkin, 1951: Cold Spring Harbor Sympos. Quant. Biol. **16**, 357.
[4] F. Kaudewitz, W. Vielmetter, und Friedrich-Freksa, 1958: Z. Naturf. **13**b, 793.
[5] M. Meselson, und Stahl, F. W., 1958: Proc. Nat. Acad. Sci. USA **44**, 671.
[6] J. D. Watson, und Crick, F. H. C., 1953: Cold Spring Harbor Sympos. Quant. Biol. **18**, 155.
[7] K. W. Mundry, und Gierer, A., 1958: Z. Vererbl. **89**, 614.
[8] J. Lederberg, und Lederberg, E. M., 1952: J. Bacteriol. **63**, 399.
[9] E. M. Witkin, 1956: Cold Spring Harbor Sympos. Quant. Biol. **21**, 123.
[10] M. Demerec, und Hanson, J., 1951: Cold Spring Harbor Sympos. Quant. Biol. **16**, 215.
[11] E. A. Adelberg, und Myers, J. W., 1953: J. Bacteriol. **65**, 348.

Inaktivierende und mutagene Wirkung salpetriger Säure auf Phagen

W. Vielmetter, H. Schuster, C. M. Wieder

Max-Planck-Institut für Virusforschung, Tübingen

Salpetrige Säure vermag die NH_2-Gruppen tragenden Basen der Ribonucleinsäure (RNS) des Tabakmosaikvirus (TMV) zu desaminieren, ohne die Integrität der Moleküle selbst zu zerstören. Die Desaminierung führt bei isolierter, biologisch aktiver TMV-RNS zum raschen Verlust der Infektiosität. Im Anschluß an diese Ergebnisse von Schuster und Schramm (1958) entdeckten Gierer und Mundry (1958), daß die Basendesaminierung mit salpetriger Säure für TMV nicht allein inaktivierend, sondern stark mutagen wirkt. Diese Entdeckung wurde der Ausgangspunkt für unsere Experimente an Bakteriophagen, über die ich hier berichten werde.

Phagen sind Viren, deren Nucleinsäurekomponente im Gegensatz zum TMV ausschließlich aus Desoxyribonucleinsäure (DNS) besteht. Im Hinblick auf die Vererbungsvorgänge in höheren Zellen erscheinen Phagen ein geeignetes Modell für das Studium der Mutagenese zu sein. Diesem Vorteil gegenüber dem TMV steht der methodische Nachteil, daß es bisher nicht gelang, biologisch aktive Phagen-DNS in gleicher Weise wie TMV-RNS in chemisch reiner Form zu isolieren. Eine Behandlung von Phagen mit salpetriger Säure hatte daher an ganzen Partikeln zu erfolgen.

Ein Phagenpartikel besitzt einen Kopfteil, der die von einer Proteinmembran umschlossene DNS enthält und einen Schwanzteil, ebenfalls aus Protein, der einen komplizierten Mechanismus verbirgt, welcher für die spezifische Absorption des Phagen an die Wirtszelle und das Eindringen der Nucleinsäure sorgt. In die Wirtszelle eingedrungene Phagen-DNS vermehrt sich dort, später wird neues Hüllprotein gebildet, in das die DNS verpackt wird. Am Ende des Vermehrungszyklus lysiert die Wirtszelle und entläßt eine Nachkommenschaft von mehreren hundert neuen Partikeln.

In Hinsicht auf die Interpretation der folgenden Versuche sei auf zwei bekannte Tatsachen hingewiesen: 1. Ein freies Phagenpartikel besitzt wie jedes Virus keinerlei eigenen Stoffwechsel. 2. Bei der Infektion dringt im wesentlichen nur die Nucleinsäurekomponente des Phagen in die Wirtszelle ein. Nach Hershey und Chase (1953) gelangen nur 3–5% des Proteins in die Zelle. Es wurde später gezeigt, daß diesem „inneren Protein" keine genetische Spezifität zukommt.

In unseren Versuchen behandelten wir nur freie Phagenpartikel mit salpetriger Säure. Wir verwendeten T2-Phagen, deren Wirt *Escherichia coli*, Stamm B, ist. Auf etwa 10^{12} Phagen/ml angereicherte und gereinigte Präparationen wurden 1 : 50 mit 0,25 m-Acetatpuffer von pH = 4,5 gemischt, bei 20° C inkubiert und zur Zeit $t = 0$ wurde $NaNO_2$ zugesetzt. Von Zeit zu Zeit wurden Proben entnommen und diese sofort 1:200 in Puffer von pH = 7 hineinverdünnt, um die Reaktion abzubrechen. Nach geeigneter weiterer Verdünnung wurden die Proben zusammen mit Indikatorbakterien plattiert. Nach Bebrütung der Platten erschienen in dem Bakterienrasen jeder Platte Plaques, die gezählt wurden, wobei ein Plaque einem infektiösen Partikel entspricht.

Das Ergebnis solcher Versuche zeigt, daß in dem Inkubationsgemisch eine rasche Inaktivierung einsetzt. Dieser Inaktivierungsprozeß soll nun diskutiert werden.

Durch geeignete Kontrollen ist zunächst nachzuweisen, daß bei pH = 4,5 ohne Nitrit keine Inaktivierung erfolgt. Ebenso hat Bebrüten mit konzentriertem $NaNO_2$ (1-m) bei pH = 7 keine inaktivierende Wirkung. Sekundäre Effekte auf die Indikatorbakterien durch das Nitrit, das in der Verdünnungsreihe nicht völlig entfernt ist, sind durch Zugabe entsprechender Nitritmengen in die Verdünnungsreihe der Kontrollproben auszuschließen. Die Kontrollen sagen eindeutig aus, daß die Inaktivierung allein auf der Wirkung der im Inkubationsgemisch entstehenden undissoziierten salpetrigen Säure beruht.

Beispiele für die quantitative Auswertung von Inaktivierungsversuchen an T2-Phagen zeigt die Abb. 1. Es ist die Zahl überlebender Phagen bezogen auf die Ausgangskonzentration im logarithmischen Maßstab gegen die Zeit aufgetragen. Als Zeiteinheit ist die Zeit τ gewählt, bei der die Partikelzahl auf e^{-1} der Ausgangskonzentration abgesunken ist. τ (in Min.) ist für verschiedene $NaNO_2$-Konzentrationen in der zugehörigen Tabelle vermerkt. Die Abb. 1 lehrt:

1. Die Inaktivierungsgeschwindigkeit ist der Nitritkonzentration und damit der Konzentration der salpetrigen Säure proportional.
2. Die Inaktivierung verläuft mit der Zeit exponentiell, ohne jede Anfangsverzögerung. Daraus folgt, daß die Desaminierung einer einzigen chemischen Gruppe eines Phagenpartikels seine Infektiosität zerstören kann. τ ist die Zeit, die notwendig ist, um pro Partikel im Mittel einen letalen Treffer zu erzeugen.

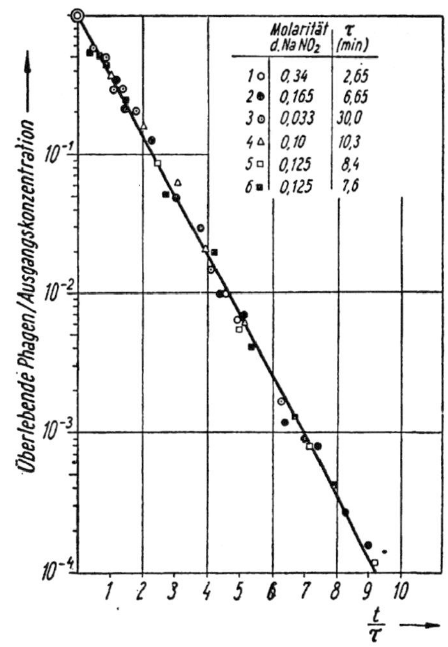

Abb. 1 (s. Text).
(Nach Vielmetter und Wieder, 1959.)

Das Versuchsergebnis enthält keinerlei Aussagen darüber, ob die letalen Treffer durch Desaminierung der DNS-Basen erfolgen. Da die NH_2-Gruppen basischer und die zyklischen Gruppen aromatischer Aminosäuren ebenfalls durch salpetrige Säure angreifbar sind, wäre es denkbar, daß Treffer im Protein den Absorptions- und Infektionsmechanismus eines Partikels zerstören. Im folgenden wird gezeigt, daß die Proteininaktivierung keinen primären Anteil am gesamten Inaktivierungsprozeß hat, ein im Hinblick auf eine mutagene Wirkung der salpetrigen Säure wichtiges Ergebnis.

Als geeignetes Modell zur Entscheidung dieser Frage bot sich der Phage PLT22 an, der sich auf dem LT2-Stamm von *Salmonella typhimurium* vermehren läßt. PLT22 ist ein temperierter Phage, der zu transduzieren vermag. Das heißt, der Phage kann Gene eines Spenderbakteriums inkorporieren und auf Empfänger übertragen. Die Übertragung der Gene durch Transduktion verlangt ebenso wie die Infektiosität einen intakten Absorptions- und Injektionsmechanismus. Werden nun die Phagen durch die vorzugsweise an der Nucleinsäurekomponente angreifenden Agentien inaktiviert (UV-Strahlung oder eingebauten ^{32}P), so findet man, daß die Transduktionsfähigkeit mehr als zehnmal langsamer inaktiviert wird, als die Infektiosität (GAREN und ZINDER, 1955). Dieser Unterschied in der Empfindlichkeit beruht darauf, daß die infektiöse Phagen-DNS um ein Vielfaches größer ist, als die transduzierten Gene.

Die relative Unempfindlichkeit der Transduktion erlaubt für PLT22 die Frage zu entscheiden, ob die inaktivierende Wirkung der salpetrigen Säure am Protein oder der DNS des Phagen angreift. Als Testphagen verwendeten wir den virulenten H5-Stamm von PLT22, der auf *Wildtyp*-Salmonellen vermehrt worden war. Zum quantitativen Nachweis der Transduktion wurden diese Phagen an Zellen einer Histidin-Mangelmutante absorbiert, die vorher mit temperierten Phagen des H1-Stammes lysogen gemacht worden war. Derart vorpräparierte Empfängerzellen sind immun, d. h. sie werden von H5 infiziert, inkorporieren die Transduktionsfragmente, überleben aber die Infektion ohne Lyse. Als Folge der Transduktion finden sich unter den Histidin$^-$-Empfängerzellen Wildtypen, die sich leicht auf Platten mit Minimalmedium entdecken lassen: Die untransduzierten Histidin$^-$-Zellen vermögen sich nicht zu vermehren, während die transduzierten Wildtypen als Kolonien aufwachsen, die gezählt werden können.

Bei den folgenden Inaktivierungsversuchen wurden H5-Phagen in gleicher Weise wie T2 mit salpetriger Säure inkubiert. Für jeden Versuchspunkt wurde nicht nur die Infektiosität, sondern auch die Transduzierfähigkeit gemessen. Das Ergebnis solcher Versuche veranschaulicht Abb. 2. Ordinate und Abszisse haben die gleichen Dimensionen wie in Abb. 1.

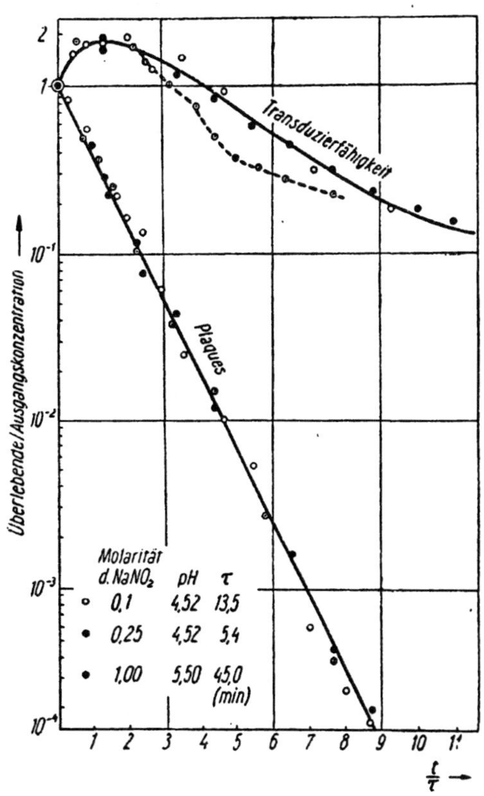

Abb. 2 (s. Text).
(Nach VIELMETTER und SCHUSTER, 1959.)

Ebenso wie bei T2 verläuft bei PLT22 die Inaktivierung der Infektiosität nach einer Eintrefferreaktion. Im Gegensatz hierzu nimmt die Transduzierfähigkeit zunächst zu, um nach Durchlaufen eines Maximums mit geringer Inaktivierungsrate abzunehmen. Diese Verhältnisse gelten für sehr verschiedene pH-Werte und Nitritkonzentrationen. Läge der Grund für die Inaktivierung an Defekten im Protein, so müßten die Kurven für Infektiosität und Transduzierfähigkeit sich decken. Die große Verschiedenheit der Kurvenverläufe beweist aber im Gegenteil, daß salpetrige Säure vorwiegend die genetische Substanz, die DNS, inaktiviert, während der im Protein liegende Absorptions- und Injektionsmechanismus ungleich weniger empfindlich ist.

Es ist nun zu fragen, ob das gleiche für T2-Phagen gilt. Hier hilft ein Vergleich der Inaktivierungsrate. Nach der Theorie ist die Trefferwahrscheinlichkeit der Größe der treffbaren Einheit und damit der Inaktivierungsrate proportional. Das Gesamtmolekulargewicht der PLT22-DNS ist fast genau dreimal kleiner als das der T2-DNS. T2-Phagen sollten daher dreimal rascher inaktiviert werden, als PLT22. Dies trifft für die

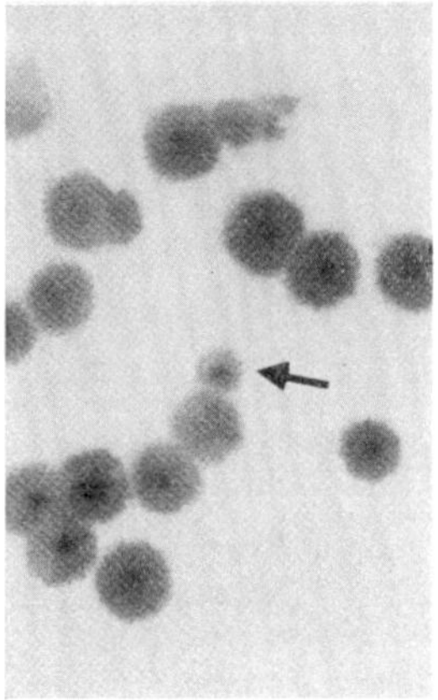

Abb. 3.

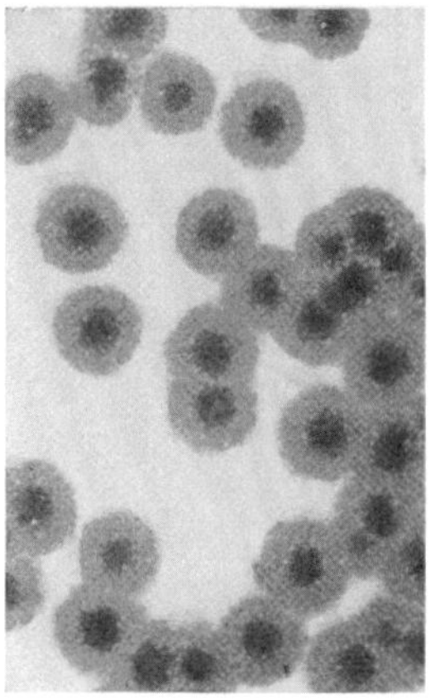

Abb. 4.

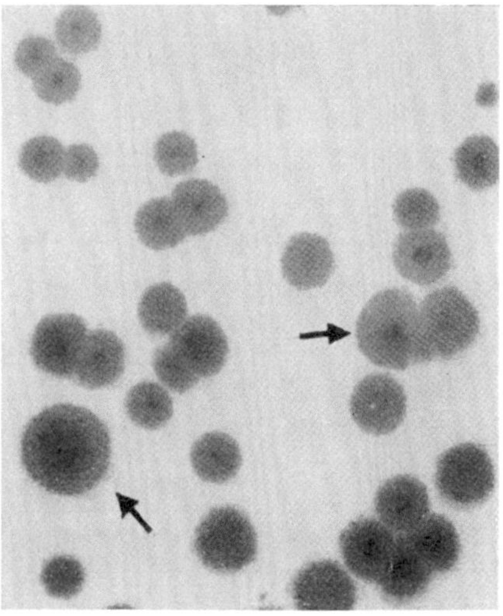

Abb. 5.

Abb. 3—5 (s. Text).

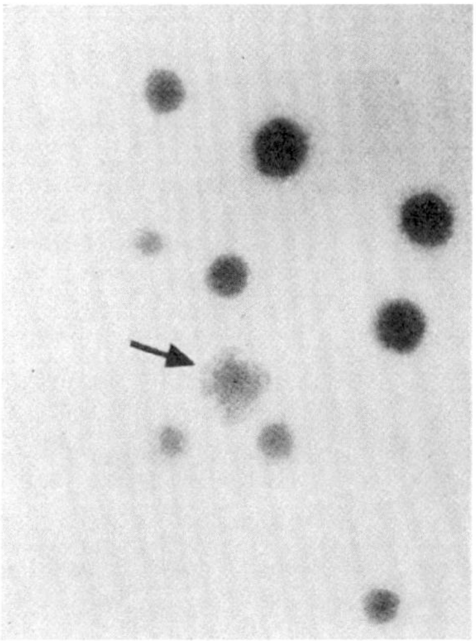

Abb. 7.

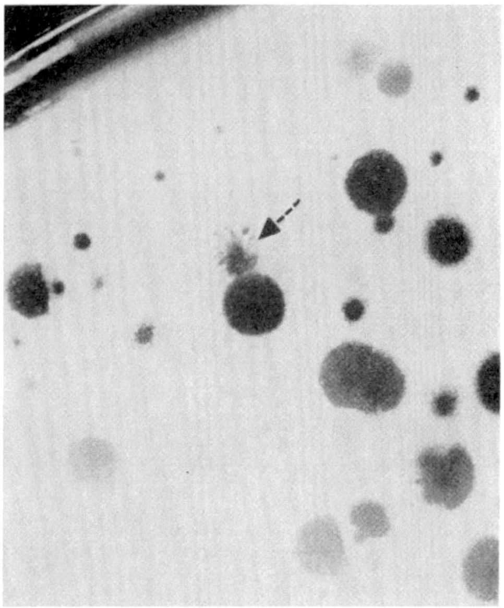

Abb. 8.

Abb. 7 und 8 (s. Text).

Inaktivierung durch Zerfall eingebauten ^{32}P auch tatsächlich zu. Die Inaktivierung von T 2 mit salpetriger Säure verläuft aber unter gleichen Bedingungen nur 1,5mal rascher als bei PLT 22. Der Unterschied um einen Faktor von 2 wird dadurch bedingt, daß pro Zeiteinheit in der T 2-DNS weniger Basen desaminiert werden. Dies folgt aus chemischen Untersuchungen der Desaminierungsreaktion, die Herr SCHUSTER bei beiden Phagen untersucht hat.

Der erste überraschende Befund war: Bei der Phagen-DNS wird im Gegensatz zur RNS nur Guanin in gleicher Rate, Cytosin viel langsamer und Adenin überhaupt nicht desaminiert. Dieser Unterschied gegenüber der RNS hängt damit zusammen, daß die NH_2-Gruppen der DNS-Basen an der Bildung von Wasserstoffbrücken beteiligt sind, die das in der von WATSON und CRICK vorgeschlagenen DNS-Struktur doppelsträngige Molekül zusammenhalten. Bei T 2-Phagen ist in der DNS an Stelle von Cytosin Hydroxy-Methyl-Cytosin vorhanden, das noch langsamer als Cytosin mit salpetriger Säure reagiert. Außerdem ist der Cytosin-Guanin-gehalt der T 2-DNS um rund 30% geringer als der PLT 22-DNS. Setzt man die quantitativen Unterschiede in Rechnung, so findet man, daß für beide Phagen die Abtötungswahrscheinlichkeit, d. h. letale Treffer/desaminierte Basen etwa 0,2 beträgt. Aus dieser Übereinstimmung schließen wir, daß, genau wie bei PLT 22, auch bei T 2 die Inaktivierung durch salpetrige Säure primär durch Basendesaminierung der DNS erfolgt. Die Tatsache, daß Veränderungen im Protein sich nur unwesentlich an der Inaktivierung beteiligen, läßt vermuten, daß sich mutagene Änderungen an der DNS in den überlebenden Partikeln tatsächlich nachweisen lassen.

Um die mutagene Wirkung salpetriger Säure direkt zu demonstrieren, wurde im wesentlichen genau wie bei den Inaktivierungsversuchen verfahren. Die verwendeten Phagenlösungen wurden durch Vermehrung jeweils einzelner T 2-Wildtyppartikel gewonnen. Bei jeder Versuchsserie wurden die schon beschriebenen Kontrollen mitgeführt. Es wurden jeweils so viele Platten mit Phagen beschickt, daß etwa 10^4 Plaques/Versuchspunkt untersucht werden konnten. Die Platten wurden 10 Stunden bebrütet und dann auf Plaques mit veränderter Morphologie hin untersucht.

Qualitativ ließen sich drei vom Wildtyp abweichende Plaque-Typen unterscheiden, die mit zunehmender Inkubationsdauer zunehmend häufiger erschienen: Nadelscharfe kleine Plaques, Plaques mit breitem trüben Rand (Abb. 3, 4) und große scharf begrenzte Plaques, die als *r*-Mutanten bekannt sind (Abb. 5). *r*-Mutanten sind auch auf dicht besäten Platten leicht zu identifizieren, weil sie von benachbarten Wildtypplaques in ihrer Ausbreitung in charakteristischer Weise gehemmt werden. Durch Passage wurde nachgewiesen, daß die morphologisch veränderten Plaque-Typen erblich stabile Eigenschaften haben.

Zur quantitativen Auswertung wurde die Zunahme der häufigeren und gut identifizierbaren *r*-Mutanten bestimmt. Abb. 6 veranschaulicht das Ergebnis. Es ist die Anzahl der

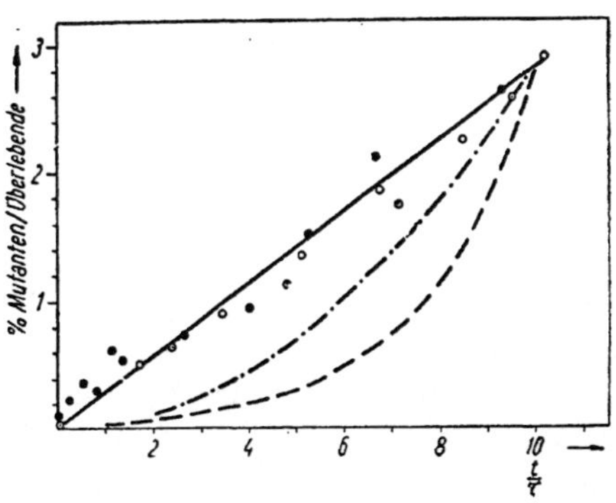

Abb. 6 (s. Text).
(Nach VIELMETTER und WIEDER, 1959).

r-Mutanten/Überlebende Wildtyp-Plaques gegen die Inkubationsdauer aufgetragen. Als Zeitmaß ist wieder die Zeit τ für die Inaktivierung gewählt. Es ist zu sehen, daß das Verhältnis Mutanten/Wildtyp von $< 0{,}05\,\%$ bis zum Versuchsende linear auf $3\,\%$ angestiegen ist, einem Zeitpunkt, bei dem die Inaktivierung auf $3 \cdot 10^{-5}$ Überlebende/Ausgangskonzentration abgesunken ist.

Der lineare Anstieg sagt aus, daß eine einzige chemische Änderung in einem Phagenpartikel eine Mutation auslösen kann. Wären zwei Treffer zur Auslösung einer Mutation notwendig, so müßte der Anstieg quadratisch erfolgen. Die in Abb. 6 (strichpunktiert) eingetragene Zweitrefferkurve weicht vom tatsächlichen Verlauf des Mutantenanstiegs so weit ab, daß ein Zweitreffervorgang mit Sicherheit auszuschließen ist.

Während der Mutantenanstieg erfolgt, findet gleichzeitig eine starke Inaktivierung statt. Es ist daher der Einwand zu prüfen, daß der Mutantenanstieg nicht auf einer Mutagenese beruht, sondern durch positive Selektion bereits vorhandener, gegen salpetrige Säure resistenter Spontanmutanten vorgetäuscht wird. Als Argument gegen eine Selektion sind zu nennen:

1. Die Inaktivierung von resistenten Spontanmutanten würde ebenso wie die der empfindlicheren Wildtypen, exponentiell mit der Zeit erfolgen, aber mit größerer Zeitkonstante τ_1. Demnach müßte sich das Verhältnis Mutanten/Wildtyp ebenfalls exponentiell ändern. Eine entsprechende Kurve für einen Selektionsvorgang dieser Art ist in Abb. 6 (unterbrochen gezeichnet) eingetragen. Ihr Verlauf ist mit den tatsächlichen Verhältnissen nicht zu vereinbaren.

2. Ein direkter Nachweis ergibt sich aus Rekonstruktionsexperimenten, bei denen durch Nitritbehandlung erzeugte Mutanten isoliert, und etwa 1 : 1 mit Wildtypphagen gemischt einer weiteren Inaktivierung unterworfen wurden. Bei einer Inaktivierung des Gemisches auf etwa 10^{-4} Überlebende/Ausgangskonzentration hätte eine Verschiebung des Verhältnisses von Mutanten/Wildtyp um einen Faktor > 100 eintreten müssen. Tatsächlich verändert sich das Verhältnis nur unbedeutend.

   Damit darf als erwiesen gelten, daß der Mutantenzuwachs auf einer mutagenen Wirkung der salpetrigen Säure beruht. Die Tatsachen des HERSHEY- und CHASE-Experimentes legen es nahe, daß alleine die Desaminierung von DNS-Basen für die Auslösung von Mutationen verantwortlich zu machen ist.

Abschließend ist zu berichten, daß eine genauere Untersuchung ergab, daß etwa die Hälfte der nach r mutierten Plaques auf den Originalplatten eingesprengte Wildtypanteile enthalten (Abb. 7). Plaques dieser Art wurden von HERSHEY (1952) nach Mischinfektion mit r^+ und r^- entdeckt und als „mottled" bezeichnet. Durch Passage kann gezeigt werden, daß mottled plaques etwa im Verhältnis 1 : 1 in reine Wildtypen und reine r-Plaques aufspalten (Abb. 8), die sich bei weiterer Passage als stabil erweisen. Dies beweist, daß es sich hier nicht um instabile Mutanten handeln kann. Die mottled plaques sind als Heterozygoten zu deuten.

Vieles spricht dafür, daß durch Mutagenese entstandene Heterozygoten durch Änderung eines Nucleotides in einem Strang der doppelsträngigen WATSON und CRICK-DNS zustande kommen. Daraus würde folgen, daß ein Einzelstrang im DNS-Molekül genetisch autonom und damit das ganze Molekül diploid ist. Wir haben gesehen, daß dies für den Inaktivierungsvorgang nicht zutreffen kann, weil er nach einer Eintreffer-Reaktion verläuft. Hierfür gibt es zahlreiche Erklärungsmöglichkeiten, deren Besprechung ich jedoch der Diskussion überlassen möchte.

Literatur

GAREN, A., und ZINDER, N. D., 1955: Virology **1**, 347.
GIERER, A., und MUNDRY, K. W., 1958: Nature (London) **182**, 1457.
HERSHEY, A. D., 1951: Cold Spring Harbor Sympos. Quant. Biol. **16**, 471.
HERSHEY, A. D., und CHASE, M., 1952: J. Gen. Physiol. **36**, 39.
MUNDRY, K. W., und GIERER, A., 1958: Z. f. Vererbungslehre **89**, 614.
SCHUSTER, H., und SCHRAMM, G., 1958: Z. Naturf. **13**b, 697.
VIELMETTER, W., und SCHUSTER, H. (in Vorbereitung).
VIELMETTER, W., und WIEDER, C. M., 1959: Z. Naturf. **14**b, 312.
WATSON, J. D., and CRICK, F. H. C., 1953: Cold Spring Harbor Sympos. Quant. Biol. **18**, 155 (1953).

Schizosaccharomyces pombe: un nouvel organisme pour l'étude de la mutagenèse chimique

H. Heslot

Laboratoire de Génétique – Institut National Agronomique, Paris 5°

1 – Problèmes non résolus

Un examen critique de l'état d'avancement des recherches dans le domaine de la mutagenèse chimique révèle que deux problèmes essentiels sont encore partiellement non résolus. Ils peuvent se définir de la façon suivante:

a – Il est nécéssaire d'établir, sur une base quantitative, quelle est la relation qui existe entre la structure moléculaire d'un composé et son activité mutagène. A la limite, la théorie devrait permettre de faire des prévisions et d'orienter la synthèse de nouveaux agents.

Présenté sous cette forme très générale, le problème a évidemment une ampleur considérable, mais celle-ci peut être provisoirement restreinte en se limitant aux substances douées du pouvoir d'alcoylation: esters sulfuriques et sulfoniques, époxides, dérivés de l'éthylène-imine, sulfures de chloro-2-éthyle, chloro-2- éthyleamines. Il s'agit d'une classe de composés relativement peu complexes dont la réactivité peut être étudiée sous l'angle des théories électroniques.

Un test quantitatif de cette nature peut concerner soit la mutation directe, soit la mutation réverse. On doit à Kølmark et Westergaard [19 à 23, 29] la mise au point d'un test bien connu de mutation réverse chez *Neurospora crassa*. Cependant ces auteurs se défendent de toute prévision quantitative car la relation qui relie le nombre de réversions pour 10^6 conidies survivantes à la concentration de l'agent mutagène est complexe et pratiquement inanalysable. Il nous semble que les courbes du type exponentiel que l'on obtient doivent résulter de ce que les conidies utilisées sont plurinucléés. – Quoi qu'il en soit, il est impossible de comparer d'une façon valable l'activité mutagène de deux substances.

En ce qui concerne la mutation directe, on doit à de Serres et Kølmark [6] un nouveau test qui fait usage des mutants rouges adenineless (ad-3) de *Neurospora crassa*. Il est encore trop tôt pour dire si ce test sera vraiment quantitatif, mais, de toute façon, *Neurospora* n'est pas un organisme idéal à tous points de vue.

b – Nous sommes encore insuffisamment renseignés sur l'existence d'une spécificité des agents chimiques mutagènes.

Dans ce domaine, il impute de distinguer nettement entre la spécificité intragénique (au niveau du site) et la spécificité intergénique (au niveau du gène).

L'existence d' une spécificité intragénique est bien établie par les travaux de KØLMARK et WESTERGAARD [19 à 23, 29], de BENZER et de FREESE [5], de FREESE [15].

Quant à la spécificité intergénique, elle apparait comme une conséquence logique de la précédente puisque les gènes sont des ensembles de sites. Cependant on peut prévoir qu'elle sera d'autant moins marquée que le nombre de sites par cistron sera plus élevé. Elle doit être à priori plus difficile à mettre en évidence que la spécificité intragénique.

Son existence ne semble pas avoir été démontrée de façon irréfutable. FAHMY et FAHMY [10 à 14], travaillant sur *Drosophila*, ont prétendu avoir mis en évidence une spécificité intergénique très nette. Non seulement ils ont affirmé que les différents agents mutagènes conduisaient à un rapport mutations visibles/mutations létales variable (0,1 dans le cas des rayons X; 0,5 dans le cas du méthane sulfonate de chloro - 2 - éthyle), mais ils ont également affirmé avoir fait muter des gènes qui ne mutaient pas avec les radiations. Ces résultats, dont le retentissement fut grand, n'ont malheureusement pas pu être confirmés par d'autres auteurs [*cf.* AUERBACH, 1 et 2] et les critiques ont été vives.

Récemment, EHRENBERG, GUSTAFSSON et LUNDQVIST [9] ont signalé que certaines mutations rares étaient plus fréquentes après traitement par l'oxyde d'éthylène ou l'éthylène-imine qu' après application des radiations ionisantes. Cependant, cela peut résulter de ce que ces agents chimiques donnent des taux de mutation beaucoup plus élevés.

Sur l'orge, les faits semblent confirmer l'existence d'une spécificité des agents chimiques mutagènes. GUSTAFSSON et EHRENBERG [7 à 9] en utilisant les mutants chlorophylliens que l'on peut classer en diverses catégories (albina, xantha, viridis etc .....) ont trouvé que la proportion de ces différentes catégories variait en fonction de l'agent mutagène utilisé.

HESLOT, FERRARY, LÉVY et MONARD [17], utilisant le même matériel et les mêmes méthodes ont trouvé des faits similaires. En particulier, il existe des différences significatives entre substances appartenant à une même série (esters sulfoniques), comme en fait foi le tableau suivant:

Agent mutagène	Nombre d'épis ségrégeant pour des mutants chlorophylliens		% d'albina
	Albina	Autres	
(1) Méthane sulfonate de méthyle $CH_3\ SO_2\ O\ CH_3$	178	298	38,1
(2) Méthane sulfonate d'éthyle $CH_3\ SO_2\ O\ CH_2\ CH_3$	68	178	27,6
(3) Méthane sulfonate de (chloro-2-éthyle) $CH_3\ SO_2\ O\ CH_2\ CH_2\ Cl$	151	216	41,1
(4) Méthane sulfonate de (bromo-2-éthyle) $CH_3\ SO_2\ O\ CH_2\ CH_2\ Br$	71	80	47,0

Comparaison $^1/_2$: $\chi^2, = 7{,}8$; $P < 0{,}01$
" $^1/_3$: $\chi^2, = 0{,}79$; $0{,}20 < P < 0{,}50$
" $^1/_4$: $\chi^2, = 3{,}75$; $0{,}05 < P < 0{,}20$
" $^2/_3$: $\chi^2, = 25{,}3$; $P \ll 0{,}01$
" $^2/_4$: $\chi^2, = 15{,}3$; $P \ll 0{,}01$
" $^3/_4$: $\chi^2, = 1{,}5$; $0{,}20 < P < 0{,}50$

Ces résultats sont en réalité d'une interprétation difficile. En admettant qu'il existe des différences significatives réelles entre les spectres, il ne s'agit que de données phénotypiques dont on ignore totalement les bases génétiques. Le matériel végétal utilisé (l'orge) ne permet pas en pratique de faire l'analyse détaillée qui serait souhaitable: nous n'avons qu'une génération par an et le nombre des gènes gouvernant la synthèse des pigments chlorophylliens est élevé. Lorsque nous observons des différences significatives dans le pourcentage d'albina avec deux agents mutagènes distincts, nous ignorons s'il s'agit de spécificité inter- ou intra-génique ou d'un mélange de ces deux effets.

2 – L'utilisation de Schizosaccharomyces pombe

Au terme de cet examen, il apparait donc que nous manquons de renseignements précis sur deux points essentiels:

Nous ne disposons pas encore d'un test biologique quantitatif qui permette de comparer entre elles les activités de différents agents:

– Nous ignorons s'il existe une spécificité inter-génique.

Pour essayer de répondre à ces deux questions, nous avons recherché un organisme approprié. Après de nombreux essais, nous avons finalement choisi la levure *Schizosaccharomyces pombe*, dont la génétique a été étudiée de façon approfondie par U. LEUPOLD, de Zürich. Pendant sa phase de croissance végétative cette levure est constituée de cellules uninucléés haploïdes se divisant par fission. Du point de vue du comportement sexuel, on distingue trois sortes de cellules h^+, h^- et h^{90}. Les types h^+ et h^- sont hétérothalliques et fusionnent en donnant un zygote diploide qui subit immédiatement la méiose avec formation de quatre ascospores à disposition linéaire. Le type h^{90} est homothallique. Les phénotypes h^+, h^- et h^{90} sont gouvernés par deux gènes étroitement liés [24 à 28].

Cette levure peut être aisément cultivée sur un milieu minimum comportant du glucose, des sels minéraux, de l'azote inorganique et d'un nombre limité de substances de croissance (ac. pantothénique, ac. nicotinique, méso-inositol, biotine).

Grâce aux travaux de U. LEUPOLD, un nombre très important de mutants biochimiques sont disponibles et les gènes distincts suivants ont été identifiés:

Adénine	: 8	Lysine	: 5	Choline	: 1
Guanine	: 2	Méthionine	: 4	Thiamine	: 1
Uracile	: 4	Leucine	: 3	Pyridoxine	: 1
Histidine	: 7	Phénylalanine	: 1		
Arginine	: 7	Ac. para amino benzoïque	: 1		

c'est à dire un total de 46 gènes représentés chacun par un ou plusieurs allèles.

Le temps de génération est court: trois jours après la copulation, les zygotes renferment des ascospores mûres. Celles-ci peuvent être analysées génétiquement par la technique classique mais fastidieuse de dissection au micromanipulateur. Cependant, au bout de quelques jours, les ascospores sont libérées spontanément par lyse de la paroi de l'asque. U. LEUPOLD a mis au point un traitement par l'alcool qui permet de tuer sélectivement les cellules végétatives, en ne conservant que les ascospores qui sont nettement plus résistantes. On peut donc procéder à des analyses génétiques en utilisant ces suspensions d'ascospores. C'est là un avantage très important par rapport à d'autres levures comme *Saccharomyces cerevisiae*.

a – Un test quantitatif de mutation réverse : Nous avons procédé à des tests systématiques de mutation réverse avec les mutants biochimiques suivants:

ad-1, ad-2, ad-3, ad-4, ad-5, ad-6, ad-7, ad-8, ur-1, ur-2, ur-3, ur-4, his-2, his-3, his-4, his-5, his-6, his-7, arg-1, arg-2, arg-3, arg-4, arg-5, arg-6, arg-7, lys-1, lys-2, lys-3, lys-4, lys-5, leuc-1, leuc-2, leuc-3. Tous ces mutants, obtenus sous l'effet de l'ultra-violet, nous avaient été donnés par U. LEUPOLD. Nous avons étudié dix allèles au locus ad-6 (tous roses) et dix allèles au locus ad-7 (tous roses). Pour tous les autres gènes mentionnés, un seul allèle a été étudié.

Sur les dix allèles ad-6 testés, un seul s'est avéré capable de mutation réverse vraie; de même sur les dix allèles ad-7, deux seulement sont susceptibles de réversion. La plupart des autres souches donnent, après traitement par les agents mutagènes, des colonies adénineless, roses capables cependant d'une croissance limitée sur le milieu minimum sans adénine.

Pour les autres gènes testés, la proportion de ceux capables de subir la mutation réverse est relativement plus élevée qu'avec ad-6 et ad-7.

Si l'on cherche quelle est la relation entre le nombre de réversions pour 10^7 cellules survivantes et la concentration de l'agent mutagène, on trouve une fonction linéaire, au moins au départ. Aux très fortes concentrations (% de survie faible) la courbe devient du type exponentiel.

A titre d'exemple, nous avons obtenu les résultats suivants avec un allèle au locus arg-1 (voir fig. 1):

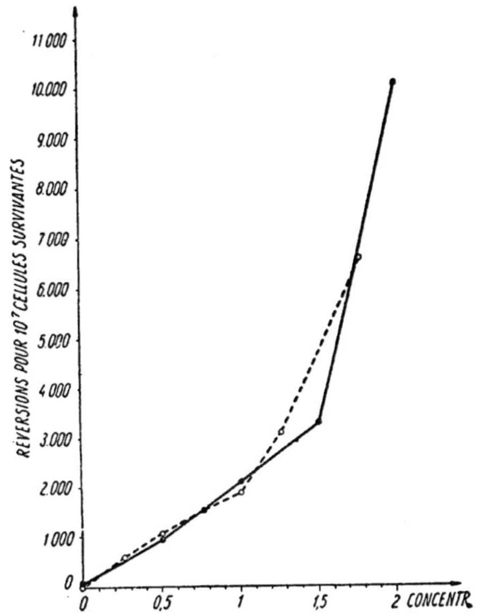

Fig. 1. Action du sulfate d'éthyle sur arg-1 (traitement: 5 heures).

1ère expérience		2e expérience	
Concentration de l'agent mutagène (Sulfate d'éthyle)	Nombre de réversions pour 10^7 cellules survivantes	Concentration de l'agent mutagène (Sulfate d'éthyle)	Nombre de réversions pour 10^7 cellules survivantes
Témoin non traité	54	Témoin non traité	6,5
0,5/500	986	0,25/500	410
1/500	2.150	0,50/500	1.028
1,5/500	3.341	0,75/500	1.463
2/500	11.125	1/500	1.903
		1,25/500	3.100
		1,75/500	6.600

Dans la première expérience, la souche n'avait pas été purifiée par réisolement avant essai, d'ou un nombre relativement élevé de réversions spontanées. Au contraire. la deuxième expérience avait été précédée d'une purification. Les deux expériences diffèrent en outre par l'âge des cultures (respectivement 5 et 3 jours) ainsi que par la température de l'essai.

Malgré ces différences, on trouve dans les deux cas la même relation linéaire au départ. Le relèvement brusque de la courbe aux hautes concentrations nous parait devoir être imputé à la présence d'amas de cellules dans la suspension traitée. De fait, pour la deuxième, nous avons trouvé la répartition suivante:

Groupes cellulaires	1	2	3	4	5	6
Fréquence	254	28	4	0	1	1

Si l'on admet qu'aux fortes doses les amas survivent de façon préférentielle, on conçoit que la relation puisse alors devenir du type exponentiel.

Il semble qu'on puisse résoudre cette difficulté en soumettant la suspension cellulaire à un traitement modéré par les ultra-sons, ce qui a pour effet de séparer les cellules sans en tuer une fraction trop importante.

Quoi qu'il en soit, l'existence d'une relation linéaire parait établie. En opérant sur une souche déterminée et dans des conditions rigoureusement standardisées, il doit donc être possible de définir pour chaque substance étudiée un indice d'activité spécifique (par exemple la concentration moléculaire conduisant à 1000 réversions pour 10^7 cellules survivantes, déduction faite des réversions spontanées). Ainsi doit-il être possible de comparer quantitativement les agents mutagènes entre eux.

Un point délicat concerne le choix du mutant de référence. Kølmark a montré de façon nette que l'action des agents mutagènes variait selon l'allèle considéré et l'on court le risque d'avoir une classification d'activité différente lorsque l'on passe d'une souche à une autre.

Dans le cas présent, ce qu'il nous intéresse de déterminer n'est pas la spécificité au niveau d'un site d'un gène particulier mais bien l'activité mutagène moyenne d'un composé, telle qu'on pourrait la mesurer en utilisant un système polygénique comparable à celui qui gouverne la formation des pigments chlorophylliens chez l'orge.

Nous pensons qu'il doit être possible de mesurer cette activité mutagène moyenne dans des expériences de mutation réverse utilisant un mutant biochimique correctement choisi. Le cistron correspondant ne doit différer du type normal que par une altération locale très minime et facilement réparable. Il doit par conséquent montrer un taux de mutation réverse spontané notable. On évitera donc d'utiliser des mutants biochimiques induits par les rayons X (qui passent pour induire fréquemment de petites délétions intra-géniques) et l'on choisira de préférence des mutants obtenus sous l'effet de l'ultra-violet. Il n'est pas impossible qu'une solution encore plus satisfaisante réside dans l'emploi de mutants induits par les antifoliques (aminoptérine, méthotrexate) dont nous avons récemment démontré l'activité mutagène.

b – Tests de spécificité inter- et intra-génique:

Pour tenter de déterminer s'il existe ou non une spécificité inter-génique, nous avons choisi la chaine de synthèse de l adénine: On sait que les huit gènes ad-1, ... , ad-8 y interviennent. Le schéma ci-dessous (fig. 2) en résume les principales étapes.

Par mutation des gènes ad-6 ou ad-7, la transformation du 5 – Aminoimidazol ribotide en acide inosique n'a plus lieu et le premier composé donne naissance à un pigment rouge qui s'accumule dans les cellules. Les mutants (ad-6)$^-$ et (ad-7)$^-$ sont par conséquent rouges.

On peut prévoir à priori – et l'expérience le confirme – la possibilité d'obtenir des mutants blancs à partir de (ad-6)$^-$ et (ad-7)$^-$, il suffit que la chaîne soit coupée une deuxième

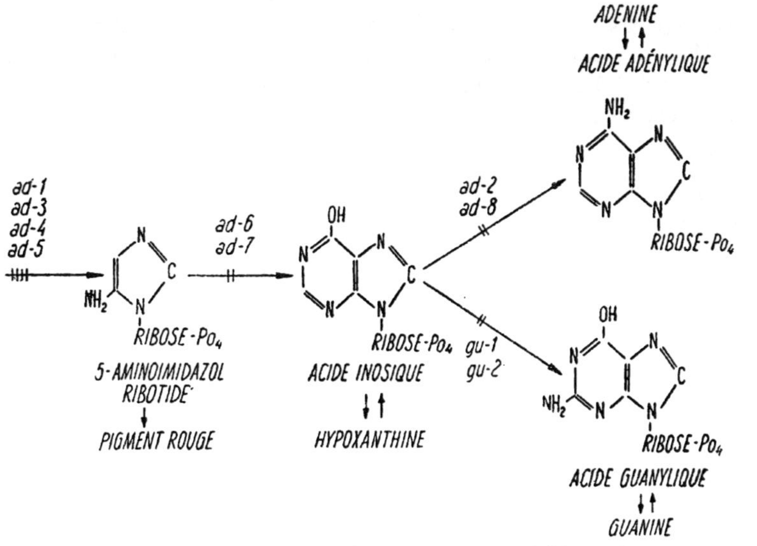

Fig. 2. Schéma simplifié de la synthèse de l'adénine chez *Schizosaccharomyces pombe*.

fois à un stade antérieur à la formation du pigment rouge. Ceci peut se produire si l'un quelconque des quatre gènes ad-1, ad-3, ad-4, ad-5 mute.

Si l'on induit dans (ad-7)⁻ une série de mutants blancs, il est possible de les analyser génétiquement et de déterminer quel est le deuxième gène mutant. On utilise pour cela la méthode des copies sur velours, initialement imaginée par J. LEDERBERG, et appliquée par U. LEUPOLD à *Schizosaccharomyces pombe*. On peut la décrire de la façon suivante: supposons que l'on dispose des quatre souches ad-1 h^{90}, ad-3 h^{90}, ad-4 h^{90} et ad-5 h^{90} qui, en raison de leur caractère homothallique, sont capables de copuler à la fois avec h⁻ et h⁺. Si l'on possède cinq mutants blancs A, B, C, D, E issus de (ad-7)⁻ et dont on désire déterminer le génotype, on opère de la façon indiquée sur la fig. 3.

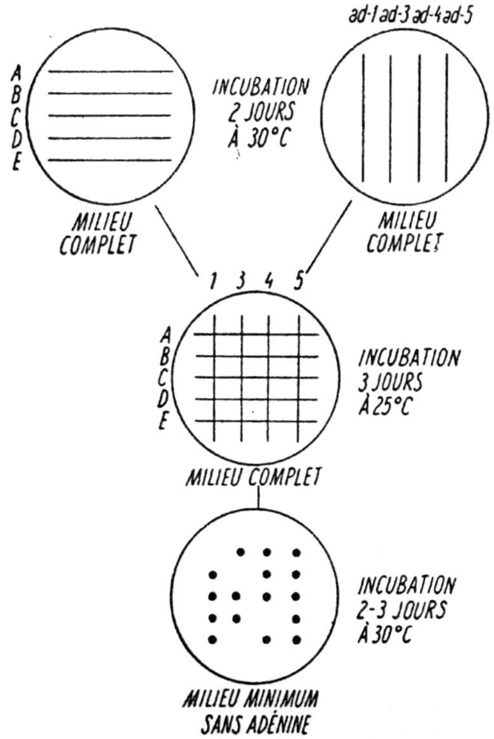

Fig. 3. Test d'allélisme de mutants adénineless chez *Schizosaccharomyces pombe*.

A la suite des copies croisées sur milieu complet, on incube les boîtes à 25° C pendant trois jours. C'est l'optimum pour la formation d'ascospores. On copie ensuite sur milieu minimum. Dans le cas présent, on voit que les souches ont la constitution:

A: ad-7 ad-1
B: ad-7 ad-3
C: ad-7 avec gène suppresseur autre
D: ad-7 ad-4
E: ad-7 ad-3

Ce type d'analyse génétique peut être appliqué à des mutants blancs, induits par des agents mutagènes variés et l'on peut étudier de cette façon s'il existe des différences dans la répartition statistique entre les quatre loci ad-1, ad-3, ad-4 et ad-5.

Nous avons trouvé les résultats suivants:

Agent Mutagène	Mutants blancs induits dans ad-7			
	ad-1	ad-3	ad-4	ad-5
1 Ultra violet	19	20	19	2
2 Sulfate d'éthyle $SO_2\ (O\ C_2\ H_5)_2$	56	84	40	4
3 Méthane sulfonate d'isopropyle $CH_3\ SO_2\ O\ CH\ (CH_3)_2$	57	41	16	4
4 Ethylène-imine 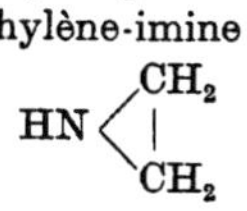	18	21	10	0

L'étude statistique donne les indications suivantes:

$$\begin{cases} \text{Comparaison } 1/2: \chi^2 = 3{,}45;\ 0{,}10 < P < 0{,}20 \\ \quad ,, \quad 1/3: \chi^2 = 8{,}8\ ;\ 0{,}01 < P < 0{,}05 \\ \quad ,, \quad 2/3: \chi^2 = 11{,}6\ ;\ P < 0{,}01 \\ \quad ,, \quad 3/4: \chi^2 = 2{,}60;\ 0{,}20 < P < 0{,}30 \\ \quad ,, \quad 1/4: \chi^2 = 1{,}99;\ 0{,}30 < P < 0{,}50 \\ \quad ,, \quad 2/4: \chi^2 = 0{,}56;\ 0{,}50 < P < 0{,}95 \end{cases}$$

Il est bien évident que le nombre de mutants analysés dans chaque cas est encore insuffisant. Il en faudrait plusieurs centaines. Il est cependant frappant de constater que les différences significatives concernent les comparaisons portant sur les effectifs les plus élevés.

Au total, les résultats ci-dessus amènent à considérer comme vraisemblable l'existence d'une spécificité intergénique des agents mutagènes.

Un point intéressant concerne le faible taux de mutation relatif du cistron ad-5. Il ne représente que 3 à 4 % du total dans tous les cas. Ceci suggère fortement que le cistron ad-5 a une structure plus simple que les trois autres et comporte un nombre de sites nettement plus faible.

Etant donné que *Schizosaccharomyces pombe* est l'un des rares organismes qui permettent l'étude de la structure fine des cistrons, il semble intéressant d'étudier la répartition des sites mutants dans ad-5 avec une série d'agents mutagènes. Ce travail doit être prochainement entrepris.

Pour l'exécution de ces recherches, nous avons disposé des moyens de travail mis à notre disposition par l'Institut National de la Recherche Agronomique.

Par ailleurs, M. Coursaget, Directeur du Service de Biologie du Commisariat à l'Energie Atomique, a bien voulu nous apporter son aide (Contrat n° 2633).

Bibliographie

[1] Auerbach, C., 1957: The study of chemical mutagens by brood pattern analysis and by the scoring of ratios between visible and lethal mutations. Z. indukt. Abst. Vererb. 88, 619–625.

[2] Auerbach, C., 1958: Mutagenic effects of alkylating agents. Ann. New York Acad. Sci. 68, 731–749.

[3] Benzer, S., 1955: Fine structure of a genetic region in bacteriophage. Proc. Nat. Acad. Sci. USA, **41**, 344–354.

[4] Benzer, S., 1957: The elementary units of heredity. The chemical basis of heredity, 70–93.

[5] Benzer, S. and Freese, E., 1958: Induction of specific mutations with 5 – bromouracil. Proc. Nat. Acad. Sci (Wash.) **44**, 112–119.

[6] de Serres, F. J. and Kölmark, H. G., 1958: A direct method for determination of forward-mutation rates in *Neurospora crassa*. Nature **182**, 1249–1250.

[7] Ehrenberg, L. and Gustafsson, A., 1957: On the mutagenic action of ethylene oxide and diepoxybutane in barley. Hereditas **43**, 595–602.

[8] Ehrenberg, L., Lundqvist, U. and Ström, G., 1958: The mutagenic action of ethylene imine in barley. Hereditas **44**, 330–336.

[9] Ehrenberg, L., Gustafsson, A. and Lundqvist, U., 1959: The mutagenic effect of ionizing radiations and reactive ethylene derivatives in barley. Hereditas **45**, 351–368.

[10] Fahmy, O. G. and Fahmy, M. J., 1954: Cytogenetic analysis of the action of carcinogens and tumour inhibitors in *Drosophila melanogaster*. II. J. of Genetics **52**, 603–619.

[11] Fahmy, O. G. and Fahmy, M. J., 1955: Cytogenetic analysis of the action of carcinogens and tumour inhibitors in *Drosophila melanogaster*. III. J. of Genetics **53**, 181–199.

[12] Fahmy, O. G. and Fahmy, M. J., 1955: Cytogenetic analysis of the action of carcinogens and tumour inhibitors in *Drosophila melanogaster*. IV. J. of Genetics **53**, 563–584.

[13] Fahmy, O. G. and Fahmy, M. J., 1956: Cytogenetic analysis of the action of carcinogens and tumour inhibitors in *Drosophila melanogaster*. V. J. of Genetics **54**, 146–164.

[14] Fahmy, O. G. and Fahmy, M. J., 1956: Mutagenicity of 2 – choroethyl methane sulfonate in *Drosophila melanogaster*. Nature **177**, 996–997.

[15] Freese, E., 1959: The difference between spontaneous and base-analogue induced mutations of phage T – 4. Proc. Nat. Acad. Sci. (Wash.) **45**, 622–623.

[16] Heslot, H. et Ferrary, R., 1958: Action génétique comparée des radiations et de quelques mutagènes sur l'orge. Annales Inst. Nat. Agronom. **44**, 133–152.

[17] Heslot, H., Ferrary, R., Levy, R. et Monard, Ch., 1959: Recherches sur les substances mutagènes (halogéno – 2 – éthyl): amines, dérivés oxygénés du sulfure de bis (chloro-2 – éthyle), esters sulfoniques et sulfuriques. C. R. Acad. Sciences **248**, 729–732.

[18] Iyer, V. N. and Szybalski, W., 1958: Two simple methods for the detection of chemical mutagens. Appl. Microbiol. **6**, 23–29.

[19] Kølmark, G. and Westergaard, M., 1949: Induced back – mutation in a specific gene of *Neurospora crassa*. Hereditas **35**, 490–506.

[20] Kølmark, G. and Westergaard, M., 1953: Further studies on chemically induced reversions at the adenine locus of *Neurospora*. Hereditas **39**, 209–224.

[21] Kølmark, G., 1953: Differential response to mutagens as studied by the *Neurospora* reverse mutation test. Hereditas **39**, 270–276.

[22] Kølmark, G., 1955: Comparative studies of monoepoxides as inducers of reverse mutations in *Neurospora*. Genetics **40**, 890–902.

[23] Kølmark, G., 1956: Mutagenic properties of certain esters of inorganic acids investigated by the *Neurospora* back – mutation test. C. R. Trav. Lab. Carlsberg. Sér. Physiol. **26**, 204–220.

[24] Leupold, U., 1950: Die Vererbung von Homothallie und Heterothallie bei *Schizosaccharomyces pombe*. C. R. Lab. Carlsberg, Sér. Physiol. **24**, 381–480.

[25] Leupold, U., 1955: Methodisches zur Genetik von *Schizosaccharomyces pombe*. Schweiz. Z. allg. Path. Bakt. **18**, 1141–1146.

[26] Leupold, U., 1955: Versuche zur genetischen Klassifizierung adenin-abhängiger Mutanten von *Schizosaccharomyces pombe*. Arch. Jul. Klaus-Stiftung Zürich **30**, 506–516.

[27] Leupold, U., 1957: Physiologisch-genetische Studien an adenin-abhängigen Mutanten von *Schizosaccharomyces pombe*. Ein Beitrag zum Problem der Pseudoallelie. Schweiz. Z. allg. Path. Bakt. **20**, 535–544.

[28] Leupold, U., 1958: Studies on recombination in *Schizosaccharomyces pombe*. Cold Spring Harbor Symp. Quant. Biol. **23**, 161–170.

[29] Westergaard, M., 1957: Chemical mutagenesis in relation to the concept of the gene. Experientia **13**, 224–234.

Action d'agents chimiques mutagènes sur quelques plantes cultivées

H. Heslot

Laboratoire de Génétique – Institut National Agronomique, Paris 5°

Notre groupe de recherches a récemment mis en évidence l'intérêt particulier du sulfate d'éthyle [1] et surtout du méthane sulfonate d'éthyle [2] pour l'induction expérimentale de mutations chez l'orge.

Si l'on prend comme mesure de l'activité mutagène le pourcentage d'épis ségrégeant pour des mutations chlorophylliennes, on a les données numériques suivantes, qui concernent l'orge de printemps de la variété Rika:

Témoin		0,05 %
Rayons γ – 25.000 r		4,5 %
Sulfate d'éthyle	: jusqu'à	14 %
Méthane sulfonate d'éthyle	: jusqu'à	50 %

On voit que le méthane sulfonate d'éthyle est un agent remarquablement efficient dont l'usage risque de transformer radicalement les conditions d'utilisation pratique de la technique des mutations pour l'amélioration des plantes cultivées.

Un phénomène intéressant sur le plan théorique est l'apparition de chimères somatiques chlorophylliennes sur les plantes directement issues de graines traitées. Ces chimères, qui se présentent comme des bandes longitudinales sur les feuilles et les tiges, sont rares après traitement par les radiations ionisantes, assez fréquentes avec le sulfate d'éthyle et très fréquentes avec le méthane sulfonate d'éthyle. Il est actuellement difficile d'en donner une interprétation génétique indiscutable car les tiges portant des chimères de quelque importance ne donnent généralement pas d'épis. Il n'a donc pas été possible d'étudier la transmission de cette anomalie chlorophyllienne en F_2. On peut penser à priori qu'il s'agit soit d'anomalies chromosomiques, soit de mutations dominantes. Enfin une action directe sur les plastes ne peut être exclue sans autre examen.

Il est frappant de constater qu'il existe une étroite corrélation entre la fréquence des chimères en F_1 et le pourcentage d'épis qui, en F_2, ségrègent pour des mutants chlorophylliens.

En France, ces nouveaux agents chimiques mutagènes ont fait l'object d'essais sur des plantes cultivées autres que l'orge, en particulier: la laitue, le blé, le riz, les arbres fruitiers, la vigne.

Je ne veux mentionner que les résultats (non encore publiés) obtenus sur la laitue par mon collègue le Dr. DOMMERGUES, de l'Institut National de la Recherche Agronomique. Ces essais, commencés en 1958, ont porté à la fois sur des variétés vertes et sur des variétés rouges (possédant des pigments anthocyaniques qui se supperposent aux pigments chlorophylliens habituels).

Les traitements ont été réalisés sur graines sèches, par trempage dans des solutions des substances mutagènes pendant une durée de 24 heures à la température du laboratoire.

Avec le sulfate d'éthyle et surtout le méthane sulfonate d'éthyle, la laitue réagit de façon très remarquable dès la F_1, par la production d'abondantes chimères chlorophylliennes et anthocyaniques, comme en fait foi le tableau suivant:

Laitue – Pourcentage de plantes X_1 présentant des chimères

1 – Essais de 1958

Variété	Agent mutagène	% de plantes X_I avec chimères
Merveille des 4 Saisons (rouge)	Rayons γ 10000r	0,086
,,	$CH_3N(CH_2CH_2Cl)_2$ 0,5‰	0,063
,,	Méthane Sulfonate de chloro-2-éthyle 1‰	0,37
,,	Oxyde d'éthylène 1‰ et 1,5‰	0,94
,,	Sulfate d'éthyle 1‰, 2‰ et 3‰	5,4

2 – Essais de 1959

Variété	Agent mutagène	Concentration de l'agent mutagène			
		1‰	2‰	3‰	4‰
Merveille des 4 Saisons (rouge)	Sulfate d' éthyle	–	4,2	4,4	8,3
Merveille des 4 Saisons (rouge)	Méthane sulfonate d'éthyle	6,9	11,5	5,65	3,3
Blonde du Prieuré (verte)	Méthane sulfonate d'éthyle	4,3	9,9	9,8	2,2

En ce qui concerne les mutations observées sur plantules en F_2, le Dr. DOMMERGUES m'a communiqué les données numériques suivantes:

Laitue – Etude des descendances X_2 – Comparaison de divers agents mutagènes

Agent mutagène	Nombre total de descendances X_2 étudiées	Nombre de descendances X_2 ségrégeant pour des mutants	%
Chlorhydrate de méthyle bis (chloro-2-éthyle) amine 0,5‰	1471	68	4,6
Méthane sulfonate de (chloro-2-éthyle) 1‰	452	18	4
Rayons γ 10000r	1028	60	6
Sulfate d'éthyle 2‰	242	79	31,6
Oxyde d'éthylène 1,5‰	233	51	22

L'oxyde d'éthylène et surtout le sulfate d'éthyle sont par conséquent des agents mutagènes puissants. Pour le méthane sulfonate d'éthyle qui a été essayé pour la première fois cette année, taux de mutation en F_2 n'est pas encore connu, mais à en juger par la fréquence des chimères chlorophylliennes et anthocyaniques en F_1 (près de 10%) on peut s'attendre à ce que le taux de mutation en F_2 soit de l'ordre de 40 à 50%.

Il est par conséquent remarquable que le méthane sulfonate d'éthyle s'avère être un agent mutagène très puissant pour des plantes aussi différentes que l'orge et la laitue. Ceci donne à penser que l'activité de cette substance pourrait bien être très générale et nous conduit à recommander l'étude de son action sur des plantes cultivées aussi variées que possible.

Bibliographie

[1] HESLOT, H., et FERRARY, R., 1958: Action génétique comparée des radiations et de quelques mutagènes sur l'orge. Annales Inst. Nat. Agronom. **44**, 133–152.

[2] HESLOT, H., FERRARY, R., LÉVY, R., et MONARD, CH., 1959: Recherches sur les substances mutagènes (halogéno-2-éthyle) amines, dérivés oxygénés du sulfure de bis (chloro-2-éthyle), esters sulfoniques et sulfuriques. C. R. Acad. Sciences, **248**, 729–732.

Note sur l'activité mutagène de certains antifoliques

H. HESLOT

Laboratoire de Génétique — Institut National Agronomique, Paris 5°

On sait quel rôle jouent l'acide folique et ses dérivés dans la biosynthèse des purines. Les deux antifoliques aminoptérine (acide 4 — amino ptéroylglutamique) et methotrexate (acide 4 — amino — N — Méthyl folique) sont de puissants inhibiteurs de la croissance cellulaire qui interfèrent avec la synthèse des bases puriques en inhibant les enzymes responsables de la réduction de l'acide folique en acide tétra hydrofolique (coenzyme F).

Nous nous sommes demandé si l'aminoptérine et la methotrexate étaient susceptibles d'induire des mutants blancs dans la souche rouge (ad — 7)⁻ de *Schizosaccharomyces pombe.*

Nous avons constaté que le traitement de cellules au repos ne conduit à aucune mutation, alors que dans ces mêmes conditions, les agents d'alcoylation en donnent des quantités importantes. Ce résultat est d'ailleurs en accord avec celui obtenu par IYER et SZYBALSKI sur *Escherichia coli* (1). Ces deux auteurs concluent que l'aminoptérine et le methotrexate sont dépourvus d' activité mutagène.

Par contre, si ces substances sont incorporées au milieu de culture à une concentration telle que la croissance demeure possible bien que ralentie, on observe la production de nombreux secteurs blancs qui, après analyse génétique, s'avèrent correspondre à des mutants doubles des types ad — 7 ad — 1, ad — 7 ad — 3, ad — 7 ad — 4 et ad — 7 ad — 5.

Pour les quelques mutants analysés, nous avons obtenu la répartition suivante:

Agent mutagène	Mutants blancs induits dans ad — 7			
	ad — 1	ad — 3	ad — 4	ad — 5
Aminoptérine	14	10	18	1
Methotrexate	8	8	14	0

Les mutants induits par ces deux antifoliques nous semblent présenter un intérêt particulier, en ce sens que la modification induite dans le gène résulte d'une »erreur métabolique« et a des chances d'être très minime et facilement réparable, d'où l'intérêt d'utiliser ces mutants dans le test quantitatif de mutation réverse auquel nous songeons.

(1) IYER, V. N. et SZYBALSKI, W., 1958: Two simple methods for the detection of chemical mutagens. Appl. Microbiol., 6, 23–29.

L'aminoptérine et le methotrexate ont été obligeamment mis à notre disposition par le Dr. J. J. DENTON, Director of organic chemical research, Lederle Laboratories, American Cyanamid — New York.

Die Wirkung einiger chemischer Mutagene auf *Xanthomonas phaseoli* var. *fuscans*.

B. Györffy

Institut für Genetik, Budapest

Die Formenmannigfaltigkeit der phytopathogenen Arten der *Xanthomonas*-Gattung, verbunden mit einer hochgradigen Wirtsspezifität, reizte uns, diese Gattung zu genetischen Untersuchungen heranzuziehen. In unserer Nachbarschaft (im Forschungsinstitut für Pflanzenschutz, Budapest) beschäftigt sich seit mehreren Jahren Herr Z. Klement mit beachtenswertem Erfolg mit den phytopathogenen *Xanthomonas*-Arten und besonders mit ihren spezifischen Bakteriophagen (Klement 1957 und 1959). Über eine genetische Bearbeitung von *Xanthomonas phaseoli* wurde bis jetzt unserer Kenntnis nach nur gelegentlich berichtet (Corey und Starr, 1958).

Wir waren zuerst bestrebt, drei Stämme (F 101, F 102 und F 11), die uns Herr Z. Klement liebenswürdigerweise zur Verfügung gestellt hat, genetisch zu markieren. Die Erzeugung biochemischer Mutanten wurde durch UV-Bestrahlung mit Anwendung der Penicillintechnik versucht. Dabei wurden bis jetzt nur einige asparagin- bzw. prolinbedürftige Mutanten erhalten. Antibiotika-resistente Varianten wurden durch hintereinandergeschaltete Überimpfungen in sukzessiv gesteigerte Streptomycin-Konzentrationen selektioniert. „Vollresistente" Formen wurden nach Ausplatten auf Streptomycin-Agar auch nach „einem Schritt" isoliert. Vorläufige Transformationsversuche mit der DNS der Streptomycin-Resistenten Varianten zeigten gewisse Erfolge (B. Györffy u. Mitarbeiter, 1959).

Seit kurzer Zeit wenden wir auch chemische Mutagene in unseren Versuchen an; es sei mir gestattet, über diese laufenden Versuche zu berichten.

Ziel unserer Untersuchungen ist es, die Mutagenität der zur Verfügung stehenden Substanzen auf *Xanthomonas phaseoli* und ihre Bakteriophagen zu prüfen und die Beeinflußbarkeit der Wirkung dieser Mutagene über den Nucleinsäurestoffwechsel zu studieren.

Die Konzentration der Mutagene und Basenanaloge, die das Wachstum nur wenig beeinträchtigt, wurde mit der spot-test-Technik von Iyer und Szybalsky (1959) bestimmt[1]). Es wurde die Wirkung der Verbindungen auf den Anteil an Streptomycin-Resistenten (gegenüber 40 γ/ml) und der Phagen-resistenten Zellen in Kulturen festgestellt, die mit Zugabe der Agenzien 23 Stunden bebrütet und dann zur Auszählung ausgeplattet wurden. Einige Ergebnisse dieser Versuche sind aus der Tabelle 1 zu ersehen.

[1]) Ich danke Frau I. Kallay, Frau S. Letenay und Fräulein M. Gabor auch an dieser Stelle für ihre gewissenhafte Mitarbeit bei der Durchführung der Experimente bzw. für die sorgfältige technische Assistenz.

Die Frequenz der Streptomycin-Resistenten war nur in den Kulturen mit Coffein, Imidazol-4,5-dicarboxylsäure oder 4-Methyl-2-thiouracil etwas höher als die der Kontrolle. Der Streptomycin-Locus wurde in Mutationsuntersuchungen bei Bakterien bereits sehr häufig untersucht (z. B. DEMEREC u. Mitarbeiter, 1951; GREER, 1958; IYER und SZYBALSKY, 1959). Die Erhöhung der Mutationsfrequenz nach mutagener Behandlung ist jedoch meistens bedeutend größer. Wir haben, wie früher bei *Serratia marcescens* (GYÖRFFY und KALLAY, 1957a, 1957b), so jetzt auch bei *Xanthomonas phaseoli* festgestellt, daß Resistente gegenüber höheren Konzentrationen von Streptomycin in einem Mutationsschritt entstehen können; diese „Vollresistenz" kann aber auch in mehreren aufeinander folgenden Schritten graduell erreicht werden (GYÖRFFY u. Mitarbeiter, 1959). Ein Abfall des Anteils an Antibiotikaresistenten bei Anwesenheit einiger Mutagene und Basenanaloge läßt eine Veränderung des Nucleinsäurestoffwechsels vermuten, wodurch die phänotypische Ausprägung der Streptomycinresistenz beeinflußt wird (PERETZ und POLGLASE, 1957).

Tabelle 1

Mutagene, Analoge	Konz. μg/ml	Zellenzahl $\times 10^8$	Strm.-res. $\times 10^{-5}$	Phag-res. $\times 10^{-2}$
Kontrolle		2,4	1,5	5,8
Äthanolamin	100	1,5	0,4	7,0
Diäthanolamin	100	1,5	0,4	3,3
Dichlorhydrin	100	3,2	1,5	8,3
Epichlorhydrin	100	4,2	0,8	3,4
Coffein	200	7,1	8,0	8,0
5-Bromuracil	100	1,9	1,6	11,0
5-Aminouracil	10	3,0	1,7	10,0
2-Thiouracil	100	2,0	0,001	9,0
4-Methyl-2-thiouracil	100	0.9	3,3	20,0
2,6-Diaminopurin	15	1,5	1,7	6,7
Benzimidazol	150	3,0	0,5	5,6
Imidazol-4,5-dicarboxylsäure	150	1,8	8,3	63,0

Phagen-resistente Varianten treten überraschenderweise in Kulturen von *Xanthomonas* mit einer so großen Häufigkeit auf (bedingt durch labile Gene?), daß diese hohe „background number" eine positive Wirkung von Mutagenen maskiert und damit eine quantitative Auswertung erschwert (WITKIN, 1947; MATNEY, SHANKEL und WYSS, 1958). Es ist deshalb fraglich, ob die hier beobachtete Zunahme an Phagen-resistenten Varianten als eine erhöhte Mutabilität oder vielleicht nur als eine positive oder negative Selektion durch die anwesenden Agenzien anzusehen ist. Es könnte ferner ebenso möglich sein, daß die erhöhte Zahl der Überlebenden entweder durch teilweise Inaktivierung bzw. verzögerte Vermehrung der Phagen, oder – wie im Falle der Streptomycin-Resistenten – durch irgendwie beeinflußte Stoffwechselbedingungen verursacht wurde.

In anderen Versuchen wurden Zellsuspensionen 2 Stunden mit den Mutagenen behandelt, dann nach entsprechenden Verdünnungen in Nährlösung suspendiert und die Zahl der Resistenten nach 0, 2, 4–5 und etwa 12 Zellgenerationen durch Ausplatten bestimmt. Es wurde weder eine Erhöhung des Anteils an Nullpunktmutationen noch an Endpunktmutationen festgestellt. Es ist hier möglich, daß die Versuchsbedingungen noch nicht restlos geeignet waren. Bekanntlich hängt die phänotypische Verzögerung der Manifestierung neu entstandener Mutationen sehr stark von den Versuchsbedingungen ab (KUBITSCHEK und BENDIGKEIT, 1958); merkwürdigerweise haben wir wiederholt in den Populationen 10 bis

14 Stunden nach der Behandlung eine Verminderung der Zahl der Phagenresistenten im Vergleich zur Kontrolle gefunden. Diese Beobachtung bedarf noch einer Erklärung. Es ist schließlich außerdem möglich, daß die von uns geprüften Mutagene auf die beiden Merkmale nicht mutagen wirken.

Die Untersuchung von zwei cytostatisch wirkenden Zuckerderivaten

Professor L. Vargha (Budapest) hat mehrere Zuckerderivate synthetisiert, die Aethylenimino- bzw. bis-Chloräthylamingruppen enthalten (Vargha, 1955). Zwei dieser Zuckerderivate, 1,2-Isopropyliden-6-äthylenimino-6-desoxy-D-Glucose (R 1) und 1,6-di- (β-Chloräthylamino)-1,6-desoxy-D-mannit-dichlorhydrat (BCM) hat uns Herr Professor Vargha freundlicherweise zur Verfügung gestellt. Die therapeutische Wirkung der Stickstofflostverbindung BCM soll die des Senfnitrogens vielfach überragen (Kellner, Nemeth und Sellei, 1955); über die Hemmwirkung von BCM auf die Antikörperbildung haben kürzlich Prof. Szentkiralyi und seine Mitarbeiter berichtet (Hadnagy u. a., 1959). BCM wird jetzt als „Degranol" durch die Firma Chinoin (Budapest) in Verkehr gebracht.

Die cytostatische Wirkung von BCM und R 1 auf Bakterien wurde im mikroskopischen Bild durch die Filamentbildung auch von uns bestätigt.

Das Wachstum der *Xanthomonas*-Population in Nährlösung wird nach einer zweistündigen Behandlung mit 1 mg/ml BCM nicht gehemmt, die gleiche Konzentration von R 1 wirkt dagegen sehr toxisch, und die Überlebenden zeigen eine etwas verminderte Wachstumsrate.

In einem Versuch wurden gewaschene Bakterien in physiologischer Kochsalzlösung (I) oder in zuckerfreier Nährlösung (II) mit Zugabe von 1 mg/ml BCM bzw. R 1 suspendiert. In bestimmten Zeitabständen wurden Proben entnommen und einerseits die Zahl der Überlebenden nach Ausplatten auf Nähragar bestimmt und andererseits gleichzeitig neue Kulturen in hundertfacher Verdünnung in Nährlösung mit Zucker hergestellt und 2 Stunden nachbebrütet (Abb. 1).

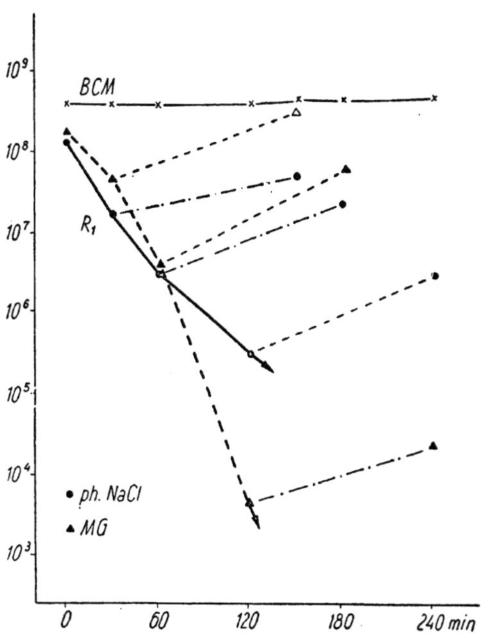

Abb. 1. Die toxische bzw. inaktivierende Wirkung von BCM und R 1 auf *Xanthomonas phaseoli*. Erklärung im Text.

Wie aus Abb. 1 zu ersehen ist, hat 1 mg/ml BCM unter den angewendeten Versuchsbedingungen keine toxische oder inaktivierende Wirkung. Bei Einwirkung von R 1 in gleicher Konzentration fällt die Zahl der Überlebenden dagegen sehr stark und linear ab. Diese toxische Wirkung von R 1 wird durch die Nachbebrütung teilweise eliminiert, und die Zellen erholen sich von der Schädigung. (Es ist zu bemerken, daß *Xanthomonas*-Zellen auch unter günstigen Bedingungen in der Log-Phase eine Generationsdauer von 2 Stunden zeigen, die aufsteigenden Kurven bedeuten also kein Wachstum!)

In einer weiteren Versuchsserie wurde die Wirkung von BCM und R 1 in Verbindung mit einer Reihe von Vorbehandlungen untersucht, die die Protein- bzw. Nucleinsäure-synthese beeinflussen. Experimente dieser Art wurden mit anderen Mutagenen (u. a. von Greer, 1958; Harold und Ziporin, 1958; Loveless und Stock, 1959) durchgeführt.

Gewaschene Zellen wurden in zuckerfreien Medien, in denen die Synthese von Protein bzw. Nucleinsäuren mit Inhibitoren aufgehoben wurde, 2 Stunden mit Belüftung bebrütet. Nach solcher Vorbehandlung wurden die Zellen mit dem Mutagen in zuckerfreier Nährlösung behandelt und danach durch Zugabe von Zucker die metabolische Aktivität der Zellen wieder hergestellt. Die Zahl der Überlebenden sowie der Anteil an Phagenresistenten wurde durch Ausplatten bestimmt.

Folgende Versuchsserien kamen zur Durchführung:

Serie 1:

Mit Zugabe von 100 γ/ml 5-Bromuracil[1]) wurde die DNS-Synthese gestört (Zamenhof, Giovanni und Greer, 1958). *Xanthomonas* benötigt Glutaminsäure zum Wachstum (Corey und Starr, 1958), deshalb wurde M-Medium (Nährsalze ohne Glutaminsäure) verwendet, um die Proteinsynthese und die Bildung von Thymin zu vermindern (Dunn und Smith, 1957; Markham, 1958; Zamenhof, Rick und Giovanni, 1958).

Serie 2:

Eine Hemmung der RNS-Synthese wurde mit 100 γ/ml 2-Thiouracil in MG Nährlösung angestrebt (Markham, 1958).

Serie 3:

Die Proteinsynthese wurde im MG-Medium (ohne Zucker) mit 1 γ/ml Chloramphenicol durch Blockieren der Konversion des labilen RNS-Aminosäure-Komplexes zu stabilen Ribonucleoprotein beeinflußt (Aronson und Spiegelman, 1958, Borek und Ryan, 1958; Gale, 1958; Harrington, 1958).

Serie 4:

Als Kontrolle diente eine Kultur mit voller MG-Nährlösung (d. h. mit Glutaminsäure- und Zuckerzusatz).

Die Ergebnisse dieser Versuche zeigten, daß Zellen, die durch zweistündige Vorbebrütung in voller Nährlösung metabolisch aktiviert waren, von BCM kaum beeinflußt wurden (Kontrolle). Die mit Chloramphenicol vorbebrüteten Zellen wurden nur wenig, die mit Bromuracil bzw. Thiouracil vorbehandelten etwas mehr von BCM geschädigt, jedoch betrug dieser Abfall der Zahl der Überlebenden nicht mehr als 5%. R 1 wirkt dagegen wesentlich ausgeprägter (Abb. 2). Die Kontrolle (R 1 in MG + Zuckernährlösung) war ebenso wie die mit BCM-Behandlung nicht inaktiviert. Zellen, die mit Chloramphenicol oder Bromuracil vorbehandelt waren, wurden von R 1 inaktiviert. Diese Schädigung konnte aber mit der Nachbehandlung (durch neue Protein bzw. RNS-Synthese) überwunden werden. Noch stärker war die R 1-Wirkung auf Zellen, in denen die RNS-

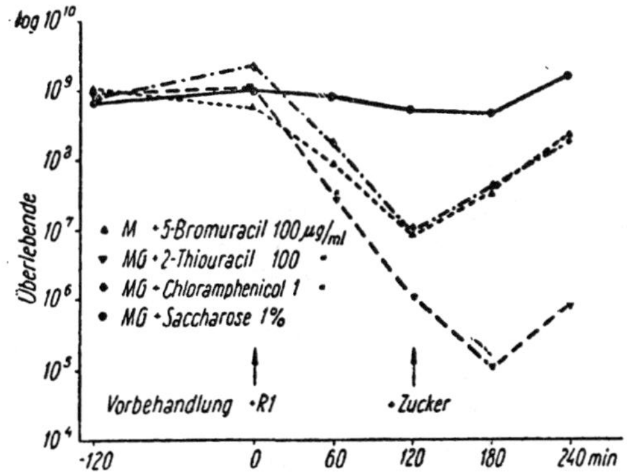

Abb. 2. Die toxische bzw. inaktivierende Wirkung von R 1 auf *Xanthomonas phaseoli* nach verschiedenartigen Vorbehandlungen. Erklärungen im Text.

[1]) Bromuracil hat uns Herr Dr. J. Skoda, Chemisches Institut der Tschechoslowakischen Akademie der Wissenschaften, Prag, liebenswürdigerweise zur Verfügung gestellt.

Synthese durch Thiouracil gestört wurde. Diese toxische Wirkung von R1 konnte durch die Nachbebrütung nicht sofort aufgehoben werden, und der Anteil revertierbarer Zellen war auch zunehmend geringer.

Über die Mutation zur Phagenresistenz haben wir besonders in dem Versuch mit BCM kein eindeutiges Bild erhalten. In den Kulturen, in denen die DNS-Synthese bzw. die Phagenvermehrung durch Bromuracil gestört sein sollte, war nach der Behandlung mit R1 die Zahl der Phagenresistenten höher als in der Kontrolle. In gleicher Weise erhöhte sich der Anteil der Phagenresistenten in den Kulturen, die mit Chloramphenicol vorbehandelt waren. Wenn jedoch die Syntheseprozesse in der Nachbehandlung durch Zuckerzusatz aktiviert wurden, fiel der Anteil der Phagenresistenten auffallend unter jenen der Kontrolle ab. Wesentlich anders verhielten sich die Kulturen mit gestörter RNS-Synthese. Der Anteil an Phagenresistenten wurde nach der Vorbehandlung mit Thiouracil vermindert, dieser Abfall wurde durch die Einwirkung von R1 noch verstärkt. Eine Nachbebrütung in zuckerhaltiger Nährlösung resultierte dann in einer plötzlichen starken Erhöhung der Zahl der Überlebenden nach Phagenbehandlung. Laufende, noch nicht abgeschlossene Versuche sollen dieses unterschiedliche Verhalten der Phagenresistenz aufklären.

Literatur

ARONSON, A. J., and SPIEGELMAN, S., 1958: On the use of chloramphenicol-inhibited systems for investigating RNA and protein synthesis. Biochim. biophys. acta 29, 214–15.

BOREK, E., und RYAN, A., 1958: Studies on a mutant of *Escherichia coli* with unbalanced ribonucleic acid synthesis. II. The concomitance of ribonucleic acid synthesis with resumed protein synthesis. J. Bacter. 75, 72–76.

COREY, R. R., and STARR, H. P., 1957: Genetic transformation of streptomycin resistance in *Xanthomonas phaseoli*. J. Bacter. 74, 146–150.

DEMEREC, M., BERTANI, G., and FLINT, J., 1951: A survey of chemicals for mutagenic action on *E. coli*. Amer. Naturalist 85, 119–136.

DUNN, D. B., and SMITH, J. D., 1957: Effects of 5-halogenated uracils on the growth of *Escherichia coli* and their incorporation into deoxyribonucleic acids. Biochemic. J. 67, 495–506.

GALE, E. F., 1958: Spezific inhibitors of protein synthesis. Symp. Soc. gen. Microbiol. 11, 212–246.

GREER, S. B., 1958: Growth inhibitors and their antagonists as mutagens and antimutagens in *Escherichia coli*. J. Gen. Microbiol. 18, 343–364.

GYÖRFFY, B., IGALI, S., KÁLLAY, I., KÁRÁSZ, I., KLEMENT, Z., und SZENDE, K., 1959: Genetic studies on a phytopathogenic *Xanthomonas*. Acta Biol. Hung. (Budapest) Suppl. 3 (im Druck).

GYÖRFFY, B., and KÁLLAY, I., 1957a: Streptomycin resistance of *Serratia marcescens*. Acta Microbiol. (Budapest) 4, 309–331.

GYÖRFFY, B., and KÁLLAY, I., 1957b: Development of resistance to streptomycin in *Serratia marcescens*. Ciba Found. Symposium on Drug Resistance in Micro-Organisms, 233–237.

HADNAGY, CS., KAPUSI, A., SZENTKIRÁLYI, E., KREPSZ, I., und SZIÁLGYI, D., 1959: Die Wirkung von antimitotischen Substanzen (Degranol, Sanamycin, Aether sulfuricus) auf die Antikörperbildung. Naturwiss. 46, 359.

HAROLD, F. M., and ZIPORIN, Z. Z., 1958: Effect of nitrogen and sulfur mustard on nucleic acid synthesis in *Escherichia coli*. Biochim. Biophys. Acta 28, 482–491.

HARRINGTON, M. G., 1958: The action of chloramphenicol on protein and nucleic acid synthesis by *Escherichia coli* strain B. J. Gen. Microbiol. 18, 767–773.

IYER, V. N., and SZYBALSKI, W., 1958: Two simple methods for the detection of chemical mutagens. Appl. Microbiol. 6, 23–29.

KELLNER, B., NÉMETH, L., SELLEI, C., 1955: Die biologische, hämatologische und geschwulsthemmende Wirkung eines neuen Stickstoff-Lost-Derivates 1,6- bis(chlor-äthylamin)-1,6-desoxy-D-mannitdichlorhydrat (BCM). Naturwiss. 4, 582–583.

KLEMENT, Z., 1957: Two new bacteriophages for bacterial pathogens of the bean. Nature **180**, 41–42.

KLEMENT, Z., and LOVAS, B., 1959: Biological and morphological characterization of the phage for *Xanthomonas phaseoli* var. *fuscans*. Növénytermelés **8**, 61–72.

KØLMARK, G., and GILES, N. H., 1955: Comparative studies of monoepoxides as inducers of reverse mutations in *Neurospora*. Genetics **40**, 890–906.

KUBITSCHEK, H. E., and BENDIGKEIT, H. E., 1958: Delay in the appearance of caffeine-induced T 5 resistance in *Escherichia coli*. Genetics **43**, 647–661.

LOVELESS, A., and STOCK, J. C., 1959: The influence of radiomimetic substances on deoxyribonucleic acid synthesis and function studied in *Escherichia coli* phage systems. I. The nature of the inactivation of T 2 phage in vitro by certain alkylating agents. Proc. Roy. Soc. (London), Ser. B. **150**, 423–445.

MARKHAM, R., 1958: Lethal synthesis. Symp. Soc. gen. Microbiol. **8**, 163–177.

MATNEY, T. S., SHANKEL, D. M., and WYSS, O., 1958: Mutations induced by ultraviolet light without attendant lethality. J. Bacteriol. **75**, 180–183.

NEWCOMBE, H. B., 1953: The delayed appearance of radiation induced genetic change in bacteria. Genetics **38**, 134–151.

PERETZ, S., and POLGLASE, W. J., 1956/1957: The action of streptomycin. Antibiot. Ann. 523–540.

VARGHA, L., 1955: Über neue Zuckerderivate mit zytostatischer Wirksamkeit. Naturwiss. **42**, 582.

WATANABE, I., KIHO, Y., and MIURA, K.-I., 1958: Effect of chloramphenicol on ribonucleic acid metabolism in *E. coli* infected with T 2 phage. Nature **181**, 1127.

WITKIN, E. M., 1947: Mutations in *Escherichia coli* induced by chemical agents. Cold Spring Harb. Symp. quant. Biol. **12**, 255–269.

ZAMENHOF, S., GIOVANNI, R., and GREER, S., 1958: Induced gene instabilization. Nature **181**, 827–829.

ZAMENHOF, S., RICK, K., und GIOVANNI, R., 1958: Further studies on the introduction of pyrimidines into deoxyribonucleic acid in *Escherichia coli*. J. Biol. Chem. **232**, 651–657.

A discussion of mutagenic specificity

C. AUERBACH and M. WESTERGAARD

1. Specificity on the "geographical" level (M. WESTERGAARD)

The possibility of a specific effect of chemical mutagens has been brought up several times during this meeting and it may be useful to introduce a discussion by attempting to formalize the concept of specificity. Mutagenic specificity may mean many different things, and it is not always quite clear which type of specificity is referred to when the phenomenon is discussed. I shall confine myself to a discussion of the problem on what may be called the "geographical" level and Dr. AUERBACH will follow up with a discussion on specificity on what she calls the "functional" level. Let me try to list the different types of specificity in the following scheme:

1. Intra-locus specificity.
   a) With respect to forward mutations ("hot spots").
   b) With respect to back mutations (allele specificity).
2. Specificity on the chromosomal level.
3. Specificity with respect to degree of damage.
4. Inter-locus specificity.

1a. Intra-locus specificity in forward mutation experiments has been well established in BENZER's system in which it has been shown that different mutagens give different "hot spots" (cf. FREESE, 1959, and my paper in this symposium).

1b. Intra-locus specificity in back mutation experiments is likewise convincingly established. It may be useful to remind you how a back mutation experiment is set up. Among the many mutants which are induced at a certain locus by mutagenic treatment, some may be unstable and undergo spontaneous reversions. These few mutants are selected for further experiments and exposed to treatment with various mutagens. It is then found that the response of each mutant allele to mutagenic treatments differs greatly, and it is possible to work out what we may call the 'back mutation pattern". Such back mutation patterns have been established for our double mutant of *Neurospora* in which we find that the back mutation pattern of the adenine-3 mutant (# 38701) is completely different from that of an inositolless mutant (# 37401). However, it has also been found that other mutants at the adenine-3 locus, or other mutants at the inositol locus concerned may show completely different patterns (KØLMARK, unpublished). Hence we are dealing with an allele-specific phenomenon and not with a locus-specific phenomenon.

Further evidence for specificity in back mutation experiments is provided by the recent observations by FREESE, who finds that the back mutation pattern of a given allele depends upon the way in which the forward mutation was induced (cf. FREESE, 1959; references in my paper, this symposium).

Undoubtedly the so-called "mutagen stable genes" in bacteria, which have been studied extensively by DEMEREC and co-workers[1]) belong to the same category, and I think that the term "mutagen stable genes" is misleading, and should be replaced by the term "mutagen stable alleles". A number of mutants in *E. coli* which will undergo spontaneous reversions do not respond to mutagenic treatments with various physical and chemical mutagens. Although most of these experiments suffer from the same shortcoming as most of WITKIN's experiments; namely, that the genetical origin of the phenomenon has not been established in crossing experiments, or in transduction or transformation experiments, I think it likely that the phenomenon is similar to the one which we find in our *Neurospora* back mutation experiments.

2. Specificity on the chromosomal level (with respect to the induction of chromosome breakage) has been established for many chemical mutagens, mainly with *Vicia faba*. Most chemicals seem to induce chromosome breakage more frequently in the heterochromatic than in the euchromatic regions. Three of the papers of this conference (by REVELL, by OCKEY, and by RIEGER and MICHAELIS) have dealt with this phenomenon.

3. Specificity with respect to degree of damage. I am now going to trespass on Dr. AUERBACH's premises because we are now actually dealing with specificity on the functional level. However, I have to mention the phenomenon here, because it may be relevant to the data to be discussed in the next paragraph. It is in my opinion quite possible that some chemicals may be 'milder' mutagens in the sense that they may inflect less severe "damage" on the genetic material than other chemical mutagens, or radiation. I use damage in quotation marks because it is misleading both from a functional and from a evolutionary point of view to speak of mutations as damage of the genetic material. However, when speaking of forward mutations in contrast to reverse mutations it is a convenient term to use.

I want to propose the hypothesis that "strong" mutagens (like ionizing radiation) may induce a relatively higher proportion of mutations of low viability (including recessive lethals) than "mild" mutagens.

4. The problem of inter-locus specificity has been rather heatedly discussed recently. It is my opinion that attempts have been made to establish this kind of specificity with organisms and with mutational systems in which the mutational analysis does not have sufficient resolving power to give a clear-cut answer.

Claims of inter-locus specificity stem from broad-spectrum mutation experiments with *Drosophila* and with barley. The *Drosophila* experiments of the FAHMY's at the Chester Beatty Institute are by far the most extensive ones. They find: 1. that the proportion of visible mutations to lethals depends upon the mutagen used and that most chemical mutagens (of the alkylating type) produce a relatively higher proportion of visibles than is produced by ionizing radiation and 2. that some chemical mutagens produce mutations at loci which have never mutated spontaneously or in X-ray experiments. Both AUERBACH and STERN have criticized some aspects of these experiments (references: C. STERN, J. Heredity *55*, pp. 276–79, 1957. O. G., and M. FAHMY. Ibid., pp. 280–87. Ch. AUERBACH, Z.I.A.V. *88*, pp. 619–25, 1957).

---

[1]) See S. W. GLOWER: "A comparative study of induced reversions in *Escherichia coli*", Carnegie Publications no. 612, 1956, pp. 121–136.

In barley both GUSTAFSSON and co-workers and HESLOT find that different chemical mutagens produce a different spectrum of chlorophyll mutations, and GUSTAFSSON finds a similar difference with respect to *erectoides* mutations.

Finally Dr. HESLOT has, likewise in this meeting, presented preliminary data obtained with his beautiful adenine system in *Schizosaccharomyces pombe*. As you will remember, the preliminary data suggest an inter-locus specificity.

Personally I am prepared to accept the experiments with *Drosophila*, barley, and yeast as suggestive but not as conclusive evidence of some kind of inter-locus specificity.

A priori one would perhaps not expect a pronounced inter-locus specificity in broad spectrum forward mutation experiments, when rather unspecific alkylating agents are used as mutagens – keeping in mind what we know, or believe we know, about the macromolecular structures of the genetical material. However, there may be two very simple explanations for the two types of specificity which have been claimed *viz.* the shift in the proportion visibles/lethals and mutations at "new loci".

The higher proportion of visibles over lethals which is obtained with some chemical mutagens as compared with X-rays in *Drosophila*, and the higher proportion of less drastic chlorophyl mutations (*e. g. viridis* and *xantha*) over the extreme albino mutations in barley may be explained by the difference between some chemical mutagens and radiation which I postulated in § 3, *viz.* that some chemical mutagens may cause less drastic mutations (and therefore mutations with less effect on viability) than ionizing radiation. This hypothesis could by tested by a conventional genetical test for allelism. In *Drosophila* this would mean testing for allelism between some of the X-ray-induced lethals and some chemically-induced visibles, and in barley, a similar test for allelism should be attempted between some of the X-ray-induced albinos against some of the chemically induced *viridis* or *xantha* mutations.

The second phenomenon, *i. e.* loci which rarely mutate spontaneously or after irradiation can be induced to mutate by certain chemicals, will now be considered. First I would like to emphasize that it would be wisest to speak of a relative specificity rather than of an absolute specificity in such cases, keeping the rather low resolving power of the mutation analysis in mind.

I would like to propose the hypothesis that we may here be dealing with an expression on the genetical level of the kind of specificity on the chromosomal level which has been well-established for chemical mutagens in *Vicia faba* (*cf.* § 2), in which it is found that many chemical mutagens of the alkylating type produce chromosome breaks preferably in heterochromatic chromosome regions. This would imply that the loci which mutate preferentially after the chemical treatments are located in/or near heterochromatic segments. In *Drosophila* there seems to be some evidence for this hypothesis (O. G. and M. FAHMY in Proc. V. Internat. Congr. Radiobiology 1956, 437–448. Oliver and Boyd, London). In barley it is unfortunately not possible to test the hypothesis, because the chromosomes are too small to allow a cytological distinction between heterochromatic and euchromatic segments.

I think it is rather important to test this very simple hypothesis, and probably more data can be obtained from *Drosophila*, which provides unique opportunities for correlating the genetical linkage map with the cytological map. It is obviously impossible to test the validity of the hypothesis on barley, but it might be worthwhile making an attempt to work out the formal genetics of *Vicia faba*, using for instance chlorophyl mutations as markers, and than trying to correlate the genetical map with the well-differentiated chromosome idiogram of that species. The third possibility for testing the hypothesis of a correlation between mutagenic specificity on the chromosomal level with inter-locus specificity would be to do more extensive work with chemical mutagens on maize in which species, surprisingly enough, very little work has been done along such lines.

Finally I should like to present a preliminary experiment which may be relevant to this discussion because it shows that two mutagens which show a very striking relative specificity in a back mutation experiment do not show a similar specificity when tested in a broad spectrum forward mutation experiment.

At an early stage of our experiments on the induction of reverse mutations in *Neurospora* we were impressed by the difference in the response of our two alleles to treatment with diepoxybutane and UV respectively. Diepoxybutane induced approximately 80 reversions per 10^6 treated macroconidia in the adenine allele, but only 2 per 10^7 reversions were induced in the inositol allele by the same treatment. UV induced approximately 3.5 reversions per 10^6 in the adenine allele, but twice as many (appr. 7.5) in the inositol allele (cf. table 2 in my paper published in this symposium).

We therefore did the following broad spectrum mutation experiment: Conidia of *Penicillium chrysogenum* were treated with UV and with meso-diepoxybutane. In each experiment the treatment was calibrated to give approximately 2 per cent surviving conidia. The conidia were plated on a complete medium and approximately 5000 colonies were picked at random from each experiment. These colonies were subsequently inspected for:

1. Morphological mutants. This group was later discarded because ist was impossible to classify the mutants unambiguously.
2. Spore colour mutants (yellow and white; the wild type had green conidia).
3. Biochemical mutants. This group was then broken down with respect to requirements.

The data are grouped in tables 1–3. In my opinion it is quite evident that the two mutagens produce almost the same spectrum of mutations. Since no obvious specificity was found by this phenotypic classification of the mutants, we did not attempt to test for allelism, the number of mutants within each group being too small. Although this experiment should be considered a preliminary one, I think it suffices to show that two mutagens which may show a striking allele-specific effect in a back mutation experiment do not show any striking specificity in a broad spectrum forward mutation experiment.

Table 1

A comparison between the mutagenic effect of ultraviolet light and of meso-diepoxybutane on *Penicillium chrysogenum.* In the UV experiment, conidia from a 6 day old culture were suspended in saline and irradiated for 60 seconds with a Hanovia "Letheray" germicidal unit at a distance of 22 cm in a quart cuvette. In the diepoxybutane experiment, conidia from a similar culture were treated with 0.11 mol meso-diepoxybutane[1]) for 43 minutes (unpublished experiments by Inger Kirk, G. Kølmark, H. Malling and M. Westergaard)

	UV	diepoxybutane
per cent survived	2.6	1.7
colonies isolated	5283	5086
colonies discarded	543	321
mutants	950	720
per cent mutants	18.0	14.2

[1]) The diepoxybutane was kindly supplied by Leo Pharmaceutical Products, Copenhagen.

Table 2

Classification of mutants

	UV	diepoxybutane
(morphological mutants *)	763	517
spore colour mutants		
white	38	16
yellow	34	15
biochemical mutants	219	198
biochemical mutants, per cent	4.1	3.9

*) discarded.

Table 3

Break-down of biochemical mutants

requirement	UV	diepoxybutane
aminoacids	139	127
vitamins	43	31
nucleic acid constituents	6	11
complex requirements	31	29
aminoacid requiring mutants		
arginine	9	10
methionine	54	52
NH_4^+	35	51 NB!
histidine	10	3
lysine	14	6
phenylalanine	1	0
leucine	6	3
proline	9	1

NB. The NH_4^+ requiring mutants were somewhat difficult to score because a rather high proportion were "leaky" mutants. The number of leaky mutants was higher after chemical treatment than after irradiation (34/4). This difference may reflect a difference between the two mutagens with respect to the degree of damage which they cause (cf. § 3, p. 117).

vitamin requiring mutants		
thiamine	14	11
choline	5	7
riboflavine	3	4
nicotinic	6	4
pyridoxine	3	1
paba	8	0 NB!
inositol	2	1
biotin	2	3

NB. The absence of paba mutants in the diepoxybutane experiment is believed to be due to an experimental error.

This is what I would have expected for theoretical reasons. I have previously compared the difference between forward mutations and back mutations by an analogy: If one throws 1000 watches hard against a wall, all the watches will probably stop, and, theoretically at least, they may have been broken in thousand different ways. It is possible, however, that some of the watches have been damaged so slightly that they may tick again if one shakes them very gently, but it would take a very gentle and "specific" kind of shaking to repair the mechanism. In the same manner there may be 1000 different ways in which a gene may be damaged by a mutagenic treatment, but we would expect that a very specific treatment would be required to repair the mutants which have not been irreversibly damaged.

2. *Specificity on the functional level* (C. Auerbach)

Professor Westergaard has analysed mutagen specificity at what one might call the geographical level. I should like to analyse it at the temporal or, better functional level. The process of mutagenesis, as I pointed out in my lecture, consists of many steps, and mutagen specificity may occur at any one of these.

(1) The mutagen must enter the cell and penetrate the nuclear membrane. Difference in penetrating ability may give rise to profound differences in mutagenic effects (e. g. between purines, as found by Kihlman), but these will be the same for the whole genome.

(2) Inside the cell, the chemical may have to change into the effective mutagen, and this derivative may be more specific than the originally used agent. Perhaps this provides the explanation of an observation by E. H. Anderson, who in *Escherichia coldi* found an oxygen effect for X-ray induced mutation at one locus, but not at another one, although the oxygen effect on survival was the same for both strains.

(3) The mutagen has to establish contact with the gene, which on our present ideas consists of DNA. But the chromosomes of all higher organisms contain protein in addition to DNA, and the two components are intimately and, probably, specifically conjoined. Since, according to Professor Westergaard's classification, I am an Evolutionist, I believe that the DNA-associated protein plays an important rôle in all processes concerning the gene. Conceivably, it may contribute to mutagen specificity. If the nucleotide sequence of DNA is reflected somehow in the amino acid sequence of the protein, then it is not difficult to imagine that the protein may sometimes selectively allow or prohibit the action of a chemical on a given gene locus. If certain amino acid mustards should, indeed, call forth specific patterns of mutation, they might do this by preferentially attaching to certain spots on the chromosome. This could be tested by synthesizing chemicals in which the same amino acid is combined with a different mutagenic group, e. g. ethylene imine. If specificity resided in the amino acid moiety, mutagenic compounds with the same amino acid group should give the same mutation pattern.

4) The mutagen reacts with the gene. This step is mutation proper; I shall call it "the primary mutation". Specificity at this level can be studied reliably only in the naked RNA of tobacco mosaic virus or the naked DNA of transforming principle, with bacteriophage in vitro as runner-up. As the genetic system of tobacco mosaic virus is still so little explored it would be of quite exceptional interest to study mutagen specificity on transforming principle from a genetically well analysed bacterial strain. It is true that mutations have not so far been induced in transforming principle; but to my knowledge there has as yet been no

systematic attempt to screen for them by, e. g., exposing transforming principle from a streptomycin sensitive strain to mutagen and screening the recipient strain for resistants. The effectively transformed bacteria could be recognized by simultaneously transferred linked markers in which recipient and donor differ.

(5) The product of the primary mutation often requires stabilization. This has been studied in bacteria by WITKIN, and by HAAS and DOUDNEY, and in *Paramecium* by KIMBALL. WITKIN's results clearly open the way to mutagen specificity at this level. She found that some mutations require protein synthesis for stabilization, others do not. If of two mutations in the same nucleus one, *a*, belongs in the first class, another *b* in the second, then a mutagen which inhibits protein synthesis will produce fewer *a* than *b* mutations.

(6) After treatment with chemical mutagens, there occurs delay over and above the stabilization delay, and different from it in nature. For chromosome rearrangements, this delay results in specific discrimination against interchromosomal as compared with intrachromosomal changes. If, as I had suggested, mutational delay after chemical treatment should be due to the formation of metastable configurations, it may well lead to mutagen specificity in experiments in which not sufficient account is taken of possible delay; for the molecular configurations formed at different loci may vary in degree of stability, depending on the mutagen used.

(7) The mutated locus has to change the biochemical pattern of the cell, and the ease with which this takes place may depend on the mutagen. GLOVER found that in bacteria the frequency of induced mutations at one locus may be influenced by the type of allele present at another locus. It is highly improbable that this interaction should take place at the level of the primary mutation. Conceivably, it occurs at the stabilization level. The most obvious possibility is that biochemically different cells offer different opportunities for the opening up of new biochemical pathways. Similarly, it seems a reasonable assumption that a chemical which produces mutation may, independently, inhibit enzymes or otherwise interfere with processes which are necessary for the formation de novo of, say, inositol in contrast to adenine. Specificity at this level would be phenotypical, resulting in the preferential induction by a given mutagen of mutations involved in the same synthetic processes. Phenotypical mutagen specificity is fairly easy to imagine in micro-organisms, where genotype and biochemical phenotype are connected by a short and, probably, direct reaction chain. It is more difficult to imagine phenotypic mutagen specificity in macro-organisms, where the phenotype is the result of complicated inter-actions at the morphological and physiological as well as the biochemical levels, and where many divisions and even fertilization intervene between the mutation and the mutated phenotype. Yet, no case of phenotypic mutagen specificity has so far been encountered in micro-organisms, whereas claims for such specificity have been made for *Drosophila* and barley. For *Drosophila*, as I discussed in my lecture, the evidence is not convincing; for barley alternative interpretations – such as different proportions of deficiencies to point mutations – seems possible.

(8) The mutated cell has to multiply into a mutated clone or organism. Specificity at this level might occur for similar reasons as at the previous one. This can, however, be tested more easily by reconstruction experiments.

Let me wind up this somewhat abstract discussion by a brief report on preliminary experiments which Dr. KØLMARK and I carried out in order to see at what level diepoxybutane (DEB) exercises its remarkable specificity in a particular *Neurospora* strain. This strain is a double requirer for inositol and adenine, and while UV produces slightly more inositol than adenine reversions, DEB has very little effect on the inositol allele, but produces exceedingly high frequencies of adenine reversions. I found it difficult to believe that such an unspecific compound should show this high degree of selectivity in its reaction with the nucleotides of the genome. It seemed more likely to me that specificity occured at one of the subsequent levels of mutagenesis. In particular, I considered the possibility that DEB in addition to producing mutations, might promote those biochemical processes which result in the priming of adenine synthesis, or inhibit those which result in the priming of inositol synthesis. I argued that, if this were true, treatment with DEB should promote the occurence of adenine reversions, independent whether these have been produced by its own action or by that of some other mutagen.

To test this, Dr. KØLMARK and I studied the production of mutations by UV in cells which immediately before or immediately after irradiation had been exposed to DEB. Pretreatment with DEB gave consistent results in many experiments: in pretreated cells UV induces many more adenine reversions than in non-pretreated ones, and there is no similar effect on the UV-induced inositol reversions. As an example I present the data from one such experiment.

Treatment	Survival %	Mutations per 10^6 survivors	
		inos.	aden.
DEB	83.5	0.1	1.6
UV	76.9	4.0	2.5
DEB + UV	67.4	4.2	18.8

The interpretation of this observation is still obscure. However, the fact that post-treatment with DEB gave no similar result makes it unlikely that my original idea was correct and suggests that specificity occurs nearer the primary mutation, perhaps at the level of stabilization.

Chemical mutagenesis: Biochemical and chemical points of view on mechanisms of action

L. EHRENBERG

Institute of Organic Chemistry and Biochemistry, University, Stockholm, Sweden

The organizers of this Symposium have honoured me by asking me to conclude the Symposium, in the present paper summarizing what I found the most important trends in the field of Chemical Mutagenesis. Since I was prevented from participating in the Symposium I have been given the opportunity to read the manuscripts of the lectures given by C. AUERBACH, Å. GUSTAFSSON, and M. WESTERGAARD. Since these papers are, themselves, very condensed reviews, from different points of view, on the subject which they summarize, I find it difficult to condense, still more, their contents. For these reasons I will chiefly restrict myself to my original manuscript, only touching certain details in the papers mentioned.

Introduction

In addition to physical agents such as the ionizing and ultraviolet radiations, many chemicals are known to be mutagenic. In principle, this property is shared by all chemicals with an ability to induce changes in the gene material, irrespective of whether such chemical changes are produced in a direct reaction of the compound with the gene material or via any one of a large number of possible chemical and biochemical indirect pathways. For this reason the number of mutagenic chemicals is very large, and it might well happen that the majority of known chemicals are in fact mutagenic. For the detection of mutagenic properties, however, a certain specificity in the action of the compound is required, which strongly restricts the number of chemicals which deserve the characterization as mutagenic. The compound must first possess an ability to penetrate into the cell, and, as far as somatic mutation is not under investigation in a multicellular organism, to reach the cells of the germ line. The specificity with respect to the mutagenic effects of the compound must further reach a certain level in relation to other effects on the cell, such as toxicity to the cytoplasm or nucleus, or the production of chromosome disturbances leading to the death of the cell and in consequence to the elimination of the mutated gene. – It seems practical to speak about the efficiency of a mutagenic chemical (and mutagenic agent in general) as an expression for this specificity, which might have many causes. The word effectiveness should be reserved for cases where the effect can be related to the dose, as is done in radiation experiments. In studies of the action of chemicals it is more difficult to define the "dose", although in simple systems the effect can be shown to depend on concentration times duration of the treatment (REVELL, 1952). The efficiency of a mutagen is measured as the highest mutation rate obtainable. When the curve of mutation rate versus dose (for chemi-

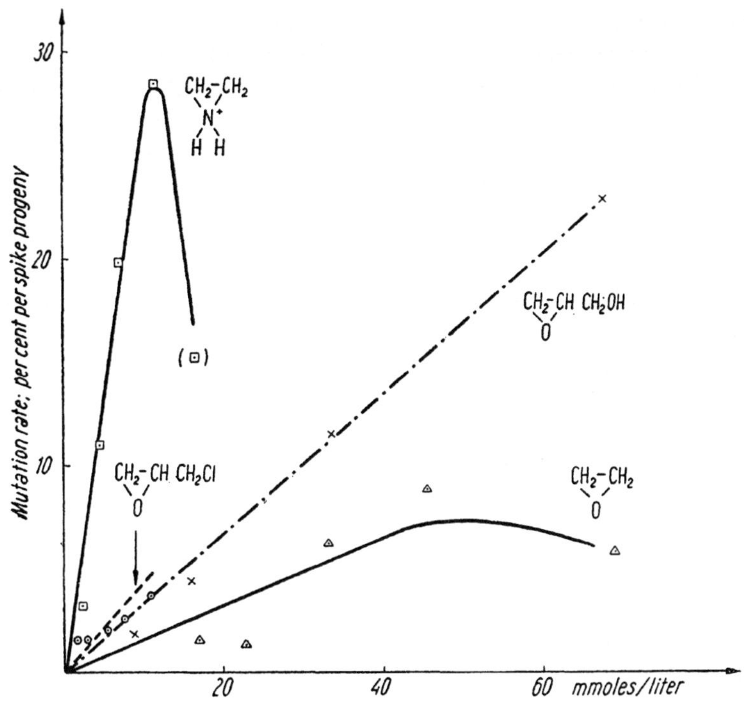

Fig. 1. Mutation rates in barley (chlorophyll mutations) as a function of molarity of solutions. Parallel experiments for determination of mutagenicity of ethyleneimine (6 h treatment), glycidol (5 h), epichlorohydrin (5 h) and ethylene oxide (10 h). Treatments at 20°C; weight of solution 10 times weight of seeds. The durations of treatment are about equal and long compared to the times required for the establishment of diffusion equilibrium. The order of effectiveness can therefore be given as ethyleneimine > epichlorohydrin ~ glycidol > ethylene oxide. If the rather insoluble epichlorohydrin is excepted this order coincides with the order of efficiency.

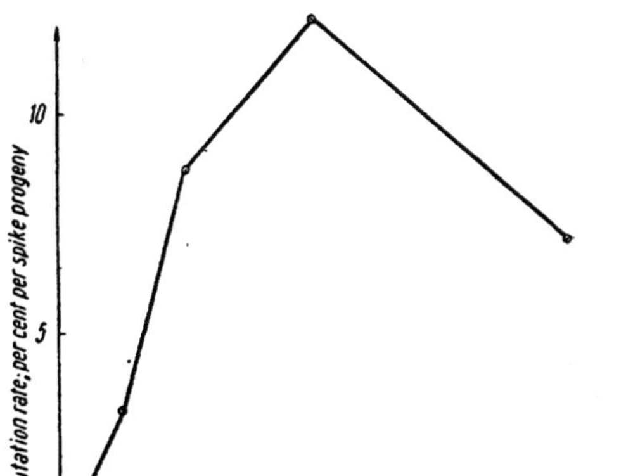

Fig. 2. Mutation rates in barley obtained with diethyl sulfate, given as a function of duration of treatment with saturated solutions, buffered to pH 6–7 by means of "tris".

cals given as concentration, duration of treatment, or their product), due to elimination of mutations at high doses, exhibits a maximum, the efficiency is identified with the latter (see Figs. 1 and 2). For epichlorohydrin (Fig. 1) the limiting factor is the solubility of the compound. – In mutation experiments the term specificity mostly refers to a differential action on individual genes, *i. e.*, to differences between mutation spectra produced by different agents.

Types of mutagenic chemicals

The mutagenic chemicals could be classified according to two different principles: according to position of the compound in the system of chemicals, or according to type of chemical reaction with biologic materials. Since, in many cases intermediates of unknown chemical constitution (e. g., in the cases of polycyclic hydrocarbons and, possibly, urethanes) seem to be the reactive agents and, in addition, the nature of the chemical reactions are not always known, a systematization with any meaning should at present be a mixture of the two principles. Only a few of the most important groups of mutagenic agents will be regarded in the following, compounds being excluded which are at present too little known to be discussed with regard to the mechanisms of mutagenesis.

Oxidants and reductants. The oxidant, iodine, was the first chemical demonstrated to induce mutation, in *Drosophila* (SACHAROFF, 1932). This activity of iodine as well as of chlorine has later been shown in barley (EHRENBERG *et al.*, 1956). The halogens are rather inefficient mutagens, *i. e.*, their effects on the biologic material are very unspecific, and it will therefore be difficult to trace more closely their action mechanism. – As for the effects of oxygen *cf.* below under "free radicals".

The mutagenic (especially chromosome breaking) effects of phenols and related compounds might be due to reactions involving a reduction. In barley a reducing agent such as the difunctional „reducton", diethyl-3,4-dihydroxy-2,5-furandicarboxylate (*cf.* EULER and HASSELQUIST, 1956) was found to induce gene mutation after treatment of resting or germinating seeds.

The mutagenic activity of the latter compound, not shared by ascorbic acid, is obviously only moderate. It cannot be excluded that the action is simply due to a state of anoxia during the germination of the seeds; soaking of barley seeds in nitrogen atmosphere has been found to lead to a small but significant increase of the mutation rate. When the latter is determined according to GUSTAFSSON's method, as frequencies of hereditary disturbances of the chlorophyll apparatus, anoxia and the "bis-reducton" give rise to similar mutation spectra, mutants of the *viridis* type being induced preferently. In this connection it should be pointed out that, in barley, many agents have been found to cause a small increase of the mutation rate in the *viridis* character. Until the mechanism of action of the agent has been established, physiologic disturbances related to those caused by anoxia cannot be excluded from being the active factor. Anaerobiosis during the soaking of seeds has also been observed to induce chromosome breakage (*cf.* LEVAN and LOTFY, 1950; MICHAELIS and RIEGER, 1958; RIEGER and MICHAELIS, 1959).

Due to its possible role as a mediator in radiation biological effects, hydrogen peroxide is the object of many mutation experiments. The compound does not seem, alone, to be able to induce mutation (KIMBALL and associates, 1955, for *Paramecium*). In combination with other compounds such as formaldehyde it may be mutagenic, however, in *Neurospora* (WESTERGAARD, 1957) or *Drosophila* (SOBELS, 1956), probably due to the production of active hydroperoxides (SOBELS, 1956; *cf.* AUERBACH, this Symposium). When H_2O_2 is given with ferric sulfate, the mixture has been shown to induce chromosome aberrations in barley (PHILIPS, 1956). In this case the equilibrium $Fe^{2+} \rightleftharpoons Fe^{3+}$ is rapidly established in the tissue and especially the Fe^{2+} ion may catalyze the decomposition of H_2O_2 into OH radicals, which cannot yet be excluded from being the active principle.

The alkylating agents form the most important group of mutagenically active chemicals. These properties were found, in the middle of the forties, independently, by RAPOPORT in the USSR for epoxides, ethyleneimine, etc., and by AUERBACH in the UK for mustard gas derivatives. About simultaneously OEHLKERS in Germany found the mutagenic activity of the urethanes; the mode of action of these compounds is still obscure – they might react *via* intermediates – but for chemical reasons it seems possible that the compounds might produce changes related to those caused by the alkylating substances. This is not a place to enter into the chemical mechanisms of reaction of these compounds, which have been summarized by several authors (e. g., by HADDOW, 1955, 1958). However, a few general principles could be mentioned.

Di- or polyfunctional alkylating agents are active at much lower concentrations than the monofunctional ones, and they are more specifically nucleotoxic in comparison with their general toxicity for the cells. This specificity is especially evident when the effect is measured as inhibition of the growth of, e. g., tumours; the inhibition seems to be due to an interaction with the chromosome material, possibly involving cross links between chromosomes or chromosome parts (GOLDACRE, LOVELESS and ROSS, 1949). When the mutagenicity of compounds belonging to this group is measured, however, much higher mutation rates can be induced by the monofunctional agents, i. e., the latter are more efficient (although often less effective) mutagens. The monofunctional compounds are active at much higher concentrations and have a much narrower concentration range before general cell toxicity sets in. Relative to cell elimination due to nucleotoxic activity they exhibit, however, a much higher mutagenic specificity. Mutation is obviously induced proportionally to concentration (EHRENBERG and GUSTAFSSON, 1957; EHRENBERG *et al.*, 1958) and might be due to one-hit changes in the gene material. Here REVELL's (1952) finding that the biologic effect of di-2,3-epoxypropyl ether depends simply on the product concentration $\times$ duration of treatment, as well as KIHLMAN's (1956) establishment of a monotonic temperature dependence and of an absence of influence of the oxygen tension during the treatment, also indicate that epoxides and compounds with a similar mode of reaction exert their action directly by chemical changes in the material, i. e. not via disturbances of the metabolism.

On the other hand sterility in barley (chiefly due to chromosome interchanges) or such effects as leaf spots in leguminous plants (probably also due to chromosome interchanges; *cf.* ZACHARIAS and EHRENBERG, data to be published), exhibit an exponential dependence on concentration, the exponent being not far from two. This has been demonstrated for ethylene oxide and ethyleneimine. Contrary to this, the diepoxybutane seems to produce sterility proportionally to the concentration (EHRENBERG and GUSTAFSSON, 1957).

The following table (1) shows the mutagenic efficiencies of monofunctional and related di- or polyfunctional alkylating substances.

At this Symposium it was pointed out by LOVELESS that, compared to the homologous alkylations, ethylation seems to involve the most specifically mutagenic change. This is indicated, *inter alia*, in comparisons in barley of the mutagenic efficiencies of the homologous alkyl esters of methanesulfonic acid (Table 1; *cf.* HESLOT *et al.*, 1959). The high mutation rates obtainable with the simplest ethylene oxide and ethyleneimine (*cf.* Table 1) might therefore be related, partly at least, to the fact that two-carbon chains, 2-hydroxyethyl and 2-aminoethyl, respectively, are introduced, whereas in the cases of the di- or polyfunctional compounds the alkyl chain contains, by necessity, four carbon atoms or more. (Also the introduction of a three-carbon group seems to be strongly mutagenic. Propyl methanesulfonate shows about half the efficiency of the corresponding ethyl ester. Compared with the fact that glycidol, the reaction of which leads to the introduction of hydroxypropyl, is the most efficient epoxide hitherto studied.) An interpretation of the factors which determine the mutagenic efficiency would require quantitative experiments with a large number of compounds. The relative importance of chain length and of the number of functional groups would no doubt be elucidated by the introduction of four-carbon mono-epoxides such as 3,4-epoxy-*n*-butanol and epoxyethyl glycol in the comparison of ethylene oxide and diepoxybutane.

Table 1

Mutagenic efficiencies of mono- and polyfunctional alkylating substances in barley. Maximum rates of mutation to disturbances of the chlorophyll apparatus, determined in the green-house test

Monofunctional compounds	Highest mutation rate, % per spike progeny	Polyfunctional compounds	Highest mutation rate, % per spike progeny
Ethylene oxide[1]) 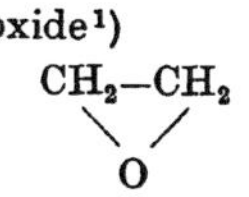	13	Diepoxybutane[1]) 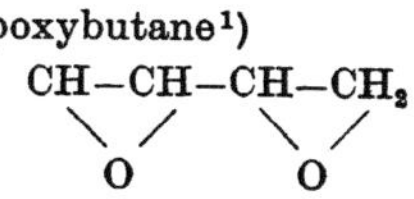	1.5
Glycidol[3]) 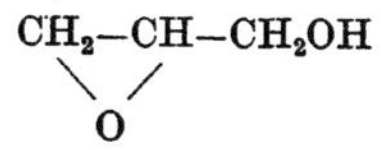	22		
Ethyleneimine[2]) 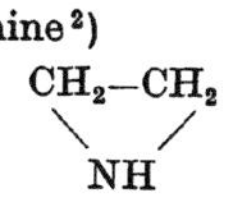	33	Triethylenemelamine[3]) 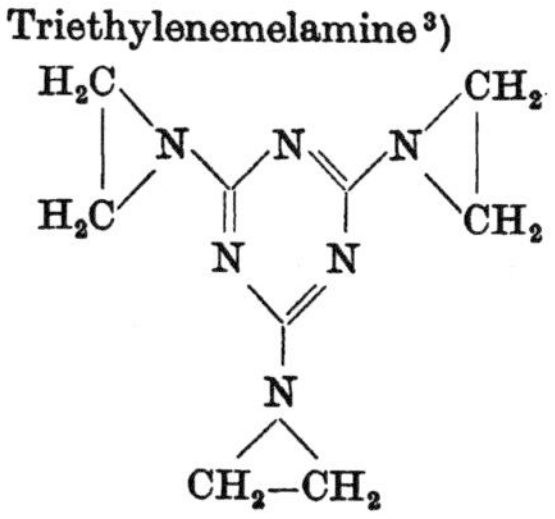	1.5
Methyl methanesulfonate[3]) $CH_3SO_2OCH_3$	16	Myleran[3]) $CH_3SO_2O(CH_2)_4OSO_2CH_3$	5.6
Ethyl methanesulfonate $CH_3SO_2OC_2H_5$	53[3]), 50[4])		
Propyl methanesulfonate[3]) $CH_3SO_2OC_3H_7$	26		

[1]) Ehrenberg and Gustafsson, 1957.
[2]) Ehrenberg *et al.*, 1958.
[3]) Ehrenberg *et al.*, unpublished.
[4]) Heslot et al., 1959.

The high mutagenic efficiency of ethylating agents has been most beautifully visualized by the fact presented by Loveless (1958 and personal communication) that it has been possible to induce mutations in phages by treatments *in vitro* by ethyl methanesulfonate or diethyl sulfate, and also ethyleneimine. Other agents, including the ionizing radiations, are well known rather to inactivate the virus particle than to induce mutation. A virus DNA molecule might roughly be taken to correspond to a gene or a locus in a cellular organism. The possibility to induce mutation in a virus would then indicate that true gene mutation, in the meaning that a changed gene is obtained, might be produced also in a cellular organism, whereas the preferential inactivation of the virus might be indicative not only of the fact that the compound in question is less specifically mutagenic but also of the possibility that its mutagenic effects are more due to losses than to changes of genes, loci, or subloci, the chemically changed DNA being incapable of reduplication.

It seems possible that the epoxides and the ethyleneimine derivatives react with negative centers such as the phosphate groups of DNA (*cf.* Elmore *et al.*, 1948; Alexander, 1952)

The fact that increasing electronegativity of substituents in the oxyrane ring increases the positive charge of the opposite carbon and leads to a higher effectiveness and probably also a higher efficiency (*cf.* Kølmark and Giles, 1955; and Fig. 1) might be explained on the basis of such assumptions. The higher specificity (efficiency) of the positively charged ethyleneiminium ion as compared to the neutral ethylene oxide might be due to the same fact (*cf.* Ehrenberg *et al.*, 1958).

Purine and pyrimidine derivatives. It can be proved that alkylating substances react with nucleic acids *in vitro*; since they, however, alkylate proteins and polysaccharides as well, the statement that the process of mutation is induced by a chemical change of the DNA has at present only the strength of a working hypothesis, which can be favoured by much argumentation. Should we assume this hypothesis to be correct (which it certainly is when mutation can be induced by *in vitro* alkylation of transforming principle DNA or infective virus nucleic acid), we do not know which are the chemical changes of the DNA, leading to mutation: alkylations of base nitrogens or of the oxygens of the "backbone" phosphates are both possible.

It has been attempted in several ways to produce a well-defined change of the base composition of the gene nucleic acids. These experiments seem at present to be able to give valuable information as to the nature of *the* chemical changes which lead to mutation.

The base composition of the nucleic acids can be changed in different ways: Purine and pyrimidine derivatives somewhat different from the natural ones may, if they are present as bases or as nucleosides or nucleotides during gene reduplication, provoke an increase of the mutation rate. Such effects have been shown for nebularine (i. e., purine-D-riboside; *cf.* Löfgren and associates 1953, 1954) in barley (Ehrenberg et al., 1956), and for 5-bromouracil or its nucleoside, desoxy-5-bromouridine, in T4 phage (Benzer and Freese, 1958). Nebularine differs from adenosine by lacking the 6-amino group, and in the desoxybromouridine the thymidine methyl is replaced by bromine. The action of such compounds can be due to an "antimetabolic" activity consisting of more or less specific disturbances of the nucleic acid synthesis, or by an incorporation of a "wrong" base into the nucleic acid. In the case of the bromouracil the latter explanation is favoured by the fact that the compound is incorporated into DNA, partly replacing the thymine, as was demonstrated already in 1952 by Weygand, Wacker and Dellweg. This is the first case described where a DNA has been produced with a well-defined chemical change. (Although it is very plausible it is not proved that this change is identical with that leading to mutation; the bromouracil which is present during the DNA synthesis may produce other disturbances of the latter.) As regards nebularine, it is not yet known whether the compound is incorporated into the gene material. Later, several other purines and pyrimidines have been found mutagenic in different materials. In barley 6-mercaptopurine, 2-aminopurine, and 8-azaguanine, as well as 5-bromouridine and especially its desoxyriboside have been found to react in similarity with nebularine, raising the mutation rate (unpublished data). – Purine and pyrimidine derivatives less closely related to those naturally occuring in the nucleic acids probably react *via* metabolic disturbances. 8-ethoxycaffeine, which causes more chromosome disturbances than gene mutation (*cf.* Ehrenberg *et al.*, 1956) has been found by Kihlman (1955) to be active *via* disturbances of oxidative phosphorylation.

The most clear-cut and perhaps only case where a well-defined chemical change of the gene nucleic acid is proved to produce mutation was recently demonstrated in the extremely beautiful work performed by the Tübingen group. *In vitro* treatment of the isolated infectious nucleic acid (in this case RNA) of tobacco mosaic virus (TMV) with nitrous acid produces, by oxidative deamination, changes *in situ*, without rupture of the polynucleotide chain, of cytosine into uracil, adenine into hypoxanthine, and guanine into xanthine (SCHUSTER and SCHRAMM, 1958). The effects were found to increase appreciably the mutation rate (GIERER and MUNDRY, 1958; MUNDRY and GIERER, 1958); the data indicate that mutation, which is a one-hit effect, can be related to the change of cytosin into uracil, which involves a change of the frequencies of the natural RNA bases, whereas the changes of the purines into "unnatural" bases produce loss of infectivity. Later, nitrous acid has also been found mutagenic in *Escherichia coli* (KAUDEWITZ, 1959) and T2 phage (VIELMETTER and WIEDER, 1959). In view of the effects on TMV, it seems plausible that also in these cases mutation was caused by changes of the bases of the gene nucleic acid, although the latter is here constituted by DNA. Since uracil (formed from cytosine) does not occur in DNA, all changes will lead to DNA-molecules which must be regarded unnatural.

By using an ingenious technique EHRENSVÄRD and associates (*cf.* 1958) could produce gene DNA with one single well-defined chemical change, a loss of the 5-methyl of thymine. After *in vivo* synthesis of bacterial DNA with thymine partly replaced by 5-^{131}I-uracil, the transforming principle (DNA) was isolated. During the subsequent nuclear decay, $^{131}I \rightarrow {}^{131}Xe$, uracil is formed in original thymine positions of otherwise intact DNA molecules. – The genetic continuation of these important experiments will no doubt fill large gaps in our knowledge of the mechanism of mutation.

Free radicals produced by the interaction of ionizing radiations with water or other inert materials of the cell have long been supposed to be the active agent in the production of genetic and other changes by radiations. At present it seems more likely that the ionizing radiations, partly at least, exert a direct action on the genetic material (*cf.* e. g., ZIMMER, 1956; ALPER, 1956). In many cases at least the potentiating effect of oxygen on radiation damage may be due to a later reaction of oxygen with unstable states of the molecules initially ionized or excited (cf. the summary given by EHRENBERG, 1959). In dry plant seeds (*Agrostis stolonifera*), where the decay of radiation induced free radicals can be followed by means of electron spin resonance, it could be proved, by the protective action of nitric oxide (NO), that the free radicals really represent those chemical states which ultimately lead to growth inhibition (SPARRMAN *et al.*, 1959). In dry barley seeds the conclusion could be extended to genetic effects, from the fact that NO decreases drastically the rates of X-ray induced chromosomal aberrations and mutations (unpublished investigations in cooperation with B. SPARRMAN).

It is at present impossible to state with certainty whether the paramagnetic centers, i. e. the free radicals, producing the effects, are located in the gene material itself (according to current conceptions the DNA) or whether they appear in neighbour molecules. For dry plant seeds data are consistent with the latter possibility, i. e. with mutation being induced by an indirect action of free radicals or their reaction products. In this connection it is of interest to know how far it is possible to induce genetic changes by treatments of living tissue with irradiated compounds, containing reactive intermediates. UV irradiated media are known to induce mutation in microorganisms (STONE *et al.*, 1947), but corresponding results for compounds exposed to ionizing radiations are still lacking. NATARAJAN and SWAMINATHAN (1958) have claimed roentgen irradiated nutrient medium to induce chromosome breakage in the root tips of wheat seedlings. In preliminary experiments we have treated barley seeds with glucose or glycine irradiated with $50 \cdot 10^6$ rad of γ-radiation (*cf.* Table 2). In the latter case an increase of the mutation rate is indicated, corresponding to that produced by irradiating the seeds with about 1000 rad. It must be pointed out, however,

Table 2

Mutagenic effect of γ-irradiated ($50 \cdot 10^6$ rad) glucose and glycine; treatment of barley seeds for 24 h immediately after the compounds had been dissolved. Controls: same concentration of unirradiated compound. Analysis of chlorophyll deficient mutants in green-house test. Resting seeds or seeds soaked for 24 h

Compound	State of seeds	Conc. %	Irradiated		Non-irradiated	
			No. of spike progenies tested	No. of spike progenies mutated	No. of spike progenies tested	No. of spike progenies mutated
Glucose	soaked	2.5	280	5	237	0
,,	,,	1.0	410	5	398	3
,,	resting	2.5	253	2	290	4
,,	,,	1.0	225	1	232	0
Glycine	soaked	1.2	439	2	432	0
,,	resting	1.2	569	5	498	1
Sum			2176	20 = 1%	2087	8 = 0.4%

p for difference between irradiated and non-irradiated compounds = 0.02* (*)
p for difference between irradiated and non-irradiated glycine = 0.05*

that these results as well as the Indian investigation need further confirmation and should be taken with due regard to a possible influence of such anoxic changes which were mentioned above. This is the more important since the mutation rates in the controls treated with unirradiated glucose and other compounds are clearly higher than the spontaneous mutation rates usually found in barley, 0.05–0.08 per cent per spike progeny. An anoxic condition could be produced by the introduction into tissue of excessive amounts of oxidizable material. – On the other hand it has been demonstrated that a water solution of glucose or other compounds previously irradiated in the crystalline state, continues for hours or days to emit light (Ahnström and Ehrenstein, 1959). This indicates that the decomposition of reactive intermediates is surprisingly slow and certainly permits the compounds to diffuse into tissue and there produce different effects. Such intermediates as aldehydes (in the case of irradiated glucose, glyceraldehyde) and hydroperoxides (*cf.* e. g. Sobels, 1958) might well possess mutagenic properties. It has been shown beyond doubt that vegetable oils may induce chromosomal rearrangements in plant seeds (Swaminathan and Natarajan, 1956). This action is probably due to reactive compounds chemically related to those produced by irradiation, i. e., hydroperoxides and aldehydes, formed during autoxidation. (In similarity with irradiated compounds, autoxidizing oils exhibit a light emission.) – Although such systems may well exist it seems less probable that, in the cases discussed here, the light as such, which is chiefly limited to the visible range, is the factor causing the biologic effects. A "radiomimetic" action of visible light seems to require the presence of a sensitizer like acridine orange; *cf.* Lüers and Borchert (1949) who presented indications of mutation in *Drosophila*, and Kihlman's (1959a, b) more conclusive work on chromosomal aberrations in *Vicia* root tips. As expressed by d'Amato, the difference between his data (1950) showing a mutagenic action of acridine orange and acriflavine in barley, and the negative results of our efforts (1956) to repeat his experiments, may be explained by the fact that the Italian sunshine, to which his material was exposed, is much more intense than the Swedish one.

The chromosome breaking and mutagenic effects of oxygen applied at supernormal pressures on pollen grains, plant seeds, or bacteria (Conger and Fairchild, 1952; Ehrenberg and J. and

M. Moutschen-Dahmen, 1957; Moutschen-Dahmen *et al.*, 1959; Fenn *et al.*, 1957; Gelin *et al.*, 1959) should perhaps be treated as an example of the reaction of oxidizing agents. It is of special interest that the breakage points induced by oxygen were found to be distributed at random within the genome. In this respect the effect of oxygen is identical with that of X-rays but different from those of other chemicals, which mostly attack certain regions of the chromosomes, e. g., heterochromatin or constrictions (Moutschen-Dahmen *et al.*, 1959). In peas oxygen and X-rays exhibit further similarity, the two mutagenic agents being of about equal efficiency and giving rise to the same spectrum of mutation types (Gelin *et al.*, *l. c.*), whereas the albina types obtained, in a very oblique spectrum, in the second generation after treatment of barley need further confirmation as to whether they are gene mutations (Moutschen-Dahmen *et al.*, *l. c.*).

It does not seem impossible that the mechanism of the action of oxygen in these respects is related chemically to autoxidation and to the enhancing effect of oxygen on radiation effects (*cf.* Ehrenberg, 1959). The radiomimetic effect demonstrated for a respiration inhibitor such as cyanide (Kihlman, 1959c) might also be explained on the basis of oxygen (and peroxides etc. formed from oxygen), the intracellular concentration of which is increased by the blocking of the centers to which it is normally bound.

The paramagnetic oxygen is in fact a double free radical, the formula of which may be written $\cdot\underline{\bar{O}}-\underline{\bar{O}}\cdot$, and it has, therefore, abilities to react with the same compounds as do other free radicals and, further, to react with radicals produced by irradiation. The (weakly mutagenic) halogens mentioned above under the heading "Oxidants" might as well be treated as free radicals. They are easily dissociated into two radicals according to $|\underline{\bar{I}}-\underline{\bar{I}}| \rightarrow |\underline{\bar{I}}\cdot + \cdot\underline{\bar{I}}|$, and iodine has in fact been used, in similarity with oxygen and nitric oxide, as a radical scavenger in radiation chemical experiments.

Specificity of mutagenic agents

It has already been mentioned that mutagenic agents, relative to their general toxic effects must possess a certain specificity with regard to their reaction with the nuclear material. In addition to this specificity, referred to as efficiency, it is clearly indicated that individual mutagenic agents or groups of mutagenic agents are specific inasmuch as they exhibit a varying affinity to different genes or parts of the gene material. This differential action is known from a wide variation of the ratio $\frac{\text{gene mutations}}{\text{chromosomal aberrations}}$, which may vary from practically zero (e. g. for ethoxycaffeine) to infinity (e. g. for nebularine), alkylating agents and ionizing radiations taking an average position (Ehrenberg *et al.*, 1956). The variation of the localization of breakage points has already been mentioned. All chemicals studied in this respect exhibit a more or less pronounced affinity to certain chromosome parts (Ford, 1949; Revell, 1953; J. and M. Moutschen-Dahmen, 1958). The agents differ also with respect to the mutation spectra obtained, e. g., as estimated from relative frequencies of phenotype groups in barley (Ehrenberg *et al.*, 1956) or *Drosophila* (Fahmy and Fahmy, 1956). We agree with the opinion presented by Auerbach at this Symposium that a specificity, established on the basis of differences between spectra of phenotypes each being produced by any one of a large number of genes, might well reflect variations in the ratio $\frac{\text{"true" gene mutations}}{\text{deficiencies}}$. However, since many data have now accumulated to indicate a clustering to certain regions of the genome of genes responsible for the production of the proteins involved in the same or related biochemical reactions (*cf.* Gustafsson's lecture at this Symposium), and since we know from the spectra of breakage points that chemicals often possess an affinity preferably for certain chromosome regions, it does not yet seem absurd to suppose that these variations in the spectra of phenotype groups might also reflect differences in mutation rates of individual genes.

We are, however, aware of the difficulties, as mentioned by AUERBACH (*cf.* EHRENBERG, 1959), which arise when we try to reconcile the enormous differences in the relative rates of back mutation in *Neurospora* (*cf.* WESTERGAARD, 1957) with the chemically rather simple structure of the polynucleotide chains or even their protein complexes. In the experiments of AUERBACH and associates it is demonstrated that in one such case the specificity is not due to differential initial affinities of the mutagenic agents to the genes but is developed early during the stabilization of the mutated gene. This extremely important experiment shows a way out of the dilemma mentioned and indicates a line along which intense research ought to be continued. It appears to us that, on the level of gross chromosomal rearrangements similar phenomena have been detected by J. and M. MOUTSCHEN-DAHMEN: In plant materials myleran produces chromosome fragments at high frequencies, whereas the number of anaphase bridges is very low, a fact indicating that exchanges are not produced. The myleran milieu seems to prevent not only exchanges between (potential) breakage points produced by myleran but also between X-ray induced breaks. It might be more than a superficial resemblance with AUERBACH's and KØLMARK's finding that the presence of diepoxybutane increases the reversion rate of an adenineless mutant, irrespectively of whether the back-mutation is produced by diepoxybutane or by ultraviolet light.

We are still far from possessing methods controlling the mutation of one single gene excluding all others. The DNA is simply built up of a varied sequence of four different nucleotides and change of one base here and there in the molecule by interaction with some active chemical, or radiation, or by incorporation of an abnormal base, might probably not be limited to an influence on one single gene. The incorporation, in organisms not permitting transformation, of oligonucleotides into the gene material could be expected to involve a more specific change. Such an incorporation requires, however, a separate control of the enzymes catalyzing the degradation of the introduced nucleotide and the incorporation of the latter into the polynucleotide chain.

Reduplication and function of the mutated gene

It has thus been established that specificities exist in the meaning that mutagenic agents differ with respect to the mutation spectra obtained. Here mutation spectra is taken in a broad sense, i. e., also including types and localisation of chromosomal aberrations. The fact that such specificities exist will be of great aid in the future clarification of how the initial chemical change of a gene becomes more or less stabilized in the mutated gene, and how the mutated gene influences the protein for the properties of which the gene is ultimately responsible.

In his lecture at this Symposium WESTERGAARD has given an excellent summary, from genetic points of view, of the essential qualifications which ought to be claimed from the experimental material in order to ensure the solution of this problem complex. His emphasizing the necessity to consider a possible difference between viruses and bacteria on one hand and higher organisms on the other with regard to the organization of the gene material, leads to the statement that qualified organisms permitting mutation experiments with sufficiently great resolving power should be sought in both groups of organisms. It is of great importance for the solution of the actual problems that it now seems possible to study "forward" mutations in higher organisms at the high resolving power previously restricted to viruses and bacteria. The investigation of mutations to adenineless in *Neurospora,* as mentioned by WESTERGAARD, and in yeast (*cf.* HESLOT's lecture) will permit a determination of the average rate of a number of mutational changes leading to adenineless. This rapid technique will therefore facilitate the comparison, with respect to efficiency and effectiveness, of long series of compounds, comparisons which are necessary for the establishments of relations between physico-chemical properties and biological action, but which cannot be performed at reasonable costs on higher plants or animals. Due to the variable response of individual mutant alleles, the otherwise excellent back-mutation technique in *Neurospora* is also of little value for such comparisons. Apart from its value for theoretical investigations, this forward mutation technique will no doubt also be extremely important for the screening of genetically active compounds which might be suspected to be hazardous to man.

In a few cases we know the chemical nature of the change produced by the mutagenic agent (*cf.* above): the change, by nitrous acid, of the cytosine of TMV nucleic acid into uracil, the change of transforming principle thymine into uracil (which case, to my knowledge, is not yet tested genetically), and the replacement of phage thymine by 5-Br-uracil.

The primary task of a continued research on mutagenic agents, including the radiations, will be to establish more generally which are the primary chemical changes caused in the gene material and later leading to mutation.

The second task will be to establish how the initially mutated gene reproduces, and the final chemical properties of the stable mutant gene. When the mutagenic agent, as in the cases mentioned, has produced a chemically changed gene nucleic acid, it will but rarely happen that the organism has the ability to copy the change when it reproduces. Only when the change, as in the case of TMV quoted, involves an alteration of the frequency or sequence of normally occurring nucleotides, might a direct copying of the change be supposed to be possible. Otherwise, the manner in which an organism reduplicates a changed template is unknown. It seems possible that many initial changes cannot be reduplicated at all. In such cases a loss of the gene or part of the certainly substructured gene, or a chromosome break, will follow. Here a third type of specificity comes into the picture. Too gross chemical rearrangements cannot be reduplicated at all and will lead to such losses. It seems possible that agents, the reactions of which lead to loss of infectivity of a virus molecule involve such gross changes. The difficulty to produce virus mutations with ionizing radiations might be an example of this. Ethylation, on the other hand by means of ethyl methanesulfonate or diethyl sulfate, in their mutagenic action on phages (LOVELESS, 1958) seem to give rise to chemical changes which permit, for some reason, the reproduction of the particle. In order to produce mutation in a plant virus by means of nitrogen mustard, the agents had to be applied during the reduplication of the virus within the host organism (LARSON *et al.*, 1955), a fact giving another aspect on the problem and probably being related to the sensitive period during DNA reduplication in a cellular organism (*cf.* REVELL, 1953). As far as back-mutation can be induced in a locus, and mutation in a supressor gene within the locus can be excluded, it is indicated that gene mutation involving a changed gene, not a loss, really exists. That a mutant gene may possess a certain type of chemical change is for the first time beautifully illustrated in BENZER's experiments with bromo-uracil, where the mutants show a rather constant reversion rate as compared with spontaneous mutants, which vary enormously in this respect.

The third part of our immediate tasks concerns the establishment of relationships between the mutated gene and the protein for the synthesis of which the gene contains, in some manner, a template. The proteins might be studied by means of sequence and content of amino acids as was first done by PAULING and, conclusively, INGRAM, for sickle cell hemoglobin in man and by the Tübingen-group for TMV protein (AACH, 1958). (In both these cases spontaneous mutants were studied.) Other important protein characteristics are antigenic properties and enzyme activities.

The establishment of the relationships between the chemically changed gene, the mutated gene, and the corresponding protein will involve the solution of fundamental questions, not only of mutation research, but it will form a solid link between genetics in general and biochemistry.

Acknowledgement. – The original data presented in this paper arise from investigations supported by the Swedish Agricultural Research Council.

References

AACH, H. G.,1958: Z. Naturforsch. **13**b, 425.
AHNSTRÖM, G., and G. VON EHRENSTEIN, 1959: Acta Chem. Scand. **13**, 855.
ALEXANDER, P., 1952: Nature **169**, 226.
ALPER, T., 1956: Radiation Research **5**, 573.
BENZER, S., and E. FREESE, 1958: Proc. Natl. Acad. Sci. U. S. **44**, 112.
CONGER, A. D., and L. M. FAIRCHILD, 1952: Proc. Natl. Acad. Sci. U. S. **38**, 289.
D'AMATO, F., 1950: Caryologia **3**, 211.
EHRENBERG, L., Å. GUSTAFSSON, and U. LUNDQVIST, 1956: Acta Chem. Scand. **10**, 492.
EHRENBERG, L., and Å. GUSTAFSSON, 1957: Hereditas **43**, 595.
EHRENBERG, L., and J. and M. MOUTSCHEN-DAHMEN, 1957: Acta Chem. Scand. **11**, 1428.
EHRENBERG, L., U. LUNDQVIST, and G. STRÖM, 1958: Hereditas **44**, 330.
EHRENBERG, L., 1959: Radiation Research Suppl. **1**, 102.
EHRENSVÄRD, G., J. LILJEKVIST, K. MOSBACH, and P. FRITZSON, 1958: Acta Chem. Scand. **12**, 485.
ELMORE, D. T., J. M. GULLAND, D. O. JORDAN, and H. F. W. TAYLOR, 1948: Biochem. J. **42**, 308.
EULER, H. VON, and H. HASSELQUIST, 1956: Hoppe-Seyler's Z. physiol. Chem. **306**, 49.
FAHMY, O. G., and M. J. FAHMY, 1956: Nature **177**, 996.
FENN, W. O., R. GERSCHMAN, D. L. GILBERT, D. E. TERWILLIGER, and F. V. COTHRAN, 1957: Proc. Natl. Acad. Sci. U. S. **43**, 1027.
FORD, C. E., 1949: Proc. 8th Intern. Cong. Genet. 1948. Suppl. Hereditas 570.
GELIN, O., L. EHRENBERG, and S. BLIXT, 1959: Agri Hortique Genetica **17**, 265.
GIERER, A., and K.-W. MUNDRY, 1958: Nature **182**, 1457.
GOLDACRE, R. J., A. LOVELESS, and W. C. J. ROSS, 1949: Nature **163**, 667.
HADDOW, A., 1959: Proc. U. N. Conf. Peaceful Uses Atomic Energy Geneva 1955 Vol. XI, p. 213.
HADDOW, A., 1958: Brit. Med. Bull. **14**, 79.
HESLOT, H., R. FERRARY, R. LÉVY, and C. MONARD, 1959: Compt. rend. Acad. Sci. **248**, 729.
KAUDEWITZ, F., 1959: Z. Naturforsch. **14**b, 528.
KIHLMAN, B. A., 1955: Exptl. Cell Research **8**, 345, 404.
– 1956: J. Biophys. Biochem. Cytol. **2**, 543.
– 1959a: Nature **183**, 976.
– 1959b: Exptl. Cell Research **17**, 590.
– 1959c: J. Biophys. Biochem. Cytol. **5**, 351.
KIMBALL, R. F., J. Z. HEARSON, and N. GAITHER, 1955: Radiation Research **3**, 435.
KØLMARK, G., and N. H. GILES, 1955: Genetics **40**, 890.
LARSON, R. H., M. A. STAHMANN, and J. D. WALKER, 1955: Proc. 2nd Conf. Potato Virus Diseases Wageningen 1954, 93.
LEVAN, A., and T. LOTFY, 1950: Hereditas **36**, 470.
LÖFGREN, N., and B. LÜNING, 1953/1954: Acta Chem. Scand. **7**, 225; *cf. ibd.* **8**, 670.
LOVELESS, A., 1958: Nature **181**, 1212.

Lüers, H., and R. Borchert, 1949: Drosophila Inform. Serv. **23**, 667.
Michaelis, A., and R. Rieger, 1958: Chromosoma **9**, 514.
Moutschen-Dahmen, J. and M., 1958: Hereditas **44**, 18, 415.
Moutschen-Dahmen, M., J. Moutschen, and L. Ehrenberg, 1959: Hereditas **45**, 230.
Mundry, K. W., and A. Gierer, 1958: Z. Vererbungslehre **89**, 614.
Natarajan, A. T., and M. S. Swaminathan, 1958: Indian J. Gen. Plant Breed. 18.
Philips, L. L., 1956: Science **124**, 889.
Revell, S. H., 1952: Thesis. University of London.
— 1953: Heredity 6 Suppl. 107.
Rieger, R. and A. Michaelis, 1959: Chromosoma **10**, 163.
Sacharoff, W., 1932: Biol. Zhur. 1:3, 1.
Schuster, H., and G. Schramm, 1958: Z. Naturforsch. **13**b, 697.
Sobels, F. H., 1956: Nature **177**, 979.
— 1958: Radiation Research **9**, 186.
Sparrman, B., L. Ehrenberg, and A. Ehrenberg, 1959: Acta Chem. Scand. **13**, 199.
Stone, W. S., O. Wyss, and F. Haas, 1947: Proc. Natl. Acad. Sci. U. S. **33**, 59.
Swaminathan, M. S., and A. T. Natarajan, 1956: Current Sci. **25**, 382.
Vielmetter, W., and C. M. Wieder, 1959: Z. Naturforsch. **14**b, 312.
Westergaard, M., 1957: Experientia **13**, 224.
Weygand, F., A. Wacker, and H. Dellweg, 1952: Z. Naturforsch. **7**b, 19.
Zacharias, M., and L. Ehrenberg, Hereditas (in press).
Zimmer, K. G., 1956: Acta Radiol. **46**, 595.

Die Kulturpflanze

Berichte und Mitteilungen aus dem Institut für Kulturpflanzenforschung der Deutschen Akademie der Wissenschaften zu Berlin in Gatersleben, Kreis Aschersleben

Herausgegeben von

RUDOLF MANSFELD, PAUL METZNER, KURT MOTHES und HANS STUBBE

Band I

1953. 227 Seiten – 11 Abbildungen im Text – 3 Pläne als Falttafeln – 9 Tafeln mit 21 Abbildungen gr. 8⁰ – DM 28,– (vergriffen)

Band II

1954. 256 Seiten – 25 Abbildungen – 14 Kunstdrucktafeln – gr. 8⁰ – DM 26,50

Band III

1955. 135 Seiten und 60 Seiten Anhang – 4 Abbildungen im Text – 10 Tafeln – 22 Tabellen – gr. 8⁰ DM 13,50 (vergriffen)

Band IV

1956. 349 Seiten – 51 Abbildungen – 49 Tabellen – 30 Tafeln – 1 Falttafel – gr. 8⁰ – DM 39,50

Band V

1957. 252 Seiten – 71 Seiten Index Seminum – 111 Abbildungen – 25 Tabellen – gr. 8⁰ – DM 34,–

Band VI

1958. 257 Seiten – 120 Abbildungen – 19 Tabellen – gr. 8⁰ – DM 35,–

Band VII

1959. 255 Seiten – 85 Seiten Index Seminum – 133 Abbildungen, davon 1 als Falttafel – 26 Tabellen gr. 8⁰ – DM 38,–

Beiheft I

Biochemie der Kulturpflanzen

Vorträge der Arbeitstagung in Gatersleben 1955
Herausgegeben von Rudolf Mansfeld und Kurt Mothes
2. unveränderte Auflage

1957. 273 Seiten – 85 Abbildungen – 14 Tafeln – gr. 8⁰ – DM 23,– (vergriffen)

Beiheft II

Vorläufiges Verzeichnis landwirtschaftlich oder gärtnerisch kultivierter Pflanzenarten

(mit Ausschluß von Zierpflanzen)
Prodromus Enumerationis Specierum Plantarum Agri- et Horticultucae
(plantis ornamentalibus exceptis)
von Rudolf Mansfeld

1959. VI, 659 Seiten – gr. 8⁰ – DM 65,–

Bestellungen durch eine Buchhandlung erbeten

AKADEMIE-VERLAG · BERLIN

HANS STUBBE

Sinn und Bedeutung der Kulturpflanzenforschung

(Vorträge und Schriften der Deutschen Akademie der Wissenschaften zu Berlin, Heft 62)

1957. 21 Seiten – 8° – DM 1,–

In diesem Vortrag auf der Feier des Leibniz-Tages der Deutschen Akademie der Wissenschaften zu Berlin am 4. 7. 1957 berichtet der Direktor des Instituts für Kulturpflanzenforschung der Deutschen Akademie der Wissenschaften und Präsident der Deutschen Akademie der Landwirtschaftswissenschaften in allgemein verständlicher Form über die vielseitigen Verflechtungen der Kulturpflanzenforschung. Sie hat entscheidenden Einfluß auf die Menschheitsgeschichte und in der Gegenwart die Aufgabe, daran mitzuarbeiten, daß die ständig wachsende Bevölkerung der Erde sich ernähren und in Glück und Wohlstand leben kann.

HANS STUBBE

Einige Ergebnisse der Mutationsforschung an Kulturpflanzen

(Sitzungsberichte der Deutschen Akademie der Wissenschaften zu Berlin, Klasse für Medizin, Jahrgang 1959, Nr. 1)

1959. 32 Seiten – 8 Text- und Tafelabbildungen – 8 Tabellen – 8° – DM 2,20

In der vorliegenden Arbeit wird gezeigt, daß sich variabel manifestierende Merkmale durch Einkreuzung in andere Sippen stabilisieren können. Offenbar wirkt die Umänderung des genotypischen Milieus fördernd auf die Ausprägung der Merkmale ein. Im zweiten Teil der Arbeit werden einige praktische Fragen der Mutationsforschung behandelt und gezeigt, daß auch züchterisch wertvolle Mutanten wie eiweißreiche Gersten und frühreife Tomaten in den Versuchen auftreten.

Bestellungen durch eine Buchhandlung erbeten

THEORIA CUM PRAXI

A K A D E M I E - V E R L A G · B E R L I N

MIX
Papier | Fördert gute Waldnutzung
FSC® C083411

Zeitfracht Medien GmbH
Ferdinand-Jühlke-Straße 7
99095 Erfurt, Deutschland
produktsicherheit@kolibri360.de